MW01675478

Nutritional Value of Grain-Based Foods

Nutritional Value of Grain-Based Foods

Special Issue Editor

Marina Carcea

MDPI • Basel • Beijing • Wuhan • Barcelona • Belgrade • Manchester • Tokyo • Cluj • Tianjin

Special Issue Editor
Marina Carcea
Council for Agricultural Research and Economics (CREA)
Italy

Editorial Office
MDPI
St. Alban-Anlage 66
4052 Basel, Switzerland

This is a reprint of articles from the Special Issue published online in the open access journal *Foods* (ISSN 2304-8158) (available at: https://www.mdpi.com/journal/foods/special_issues/nutritional_value_grain).

For citation purposes, cite each article independently as indicated on the article page online and as indicated below:

LastName, A.A.; LastName, B.B.; LastName, C.C. Article Title. *Journal Name* **Year**, *Article Number*, Page Range.

ISBN 978-3-03936-210-3 (Hbk)
ISBN 978-3-03936-211-0 (PDF)

© 2020 by the authors. Articles in this book are Open Access and distributed under the Creative Commons Attribution (CC BY) license, which allows users to download, copy and build upon published articles, as long as the author and publisher are properly credited, which ensures maximum dissemination and a wider impact of our publications.
The book as a whole is distributed by MDPI under the terms and conditions of the Creative Commons license CC BY-NC-ND.

Contents

About the Special Issue Editor

Marina Carcea is a Senior Scientist and Vice-Director of the Research Centre for Food and Nutrition of the Council for Agricultural Research and Economics (CREA) in Rome, Italy—a leading research institution under the aegis of the Italian Ministry of Agricultural, Food and Forestry Policies, where she is also a Scientific Advisor. She has over 30 years of research experience in the fields of agriculture, food, and human nutrition, which she has gained by taking part in, and coordinating, many successful research projects, involving several institutions both within Italy and abroad (Italian Ministries, EU, FAO). Dr. Carcea has authored over 200 scientific publications, many in international journals, as well as several book chapters, and has delivered lectures on her research activity at numerous congresses worldwide over the years. She has been part of multiple national and international committees regarding food and nutrition topics (Italian ministries, technological platforms, UNI, Codex Alimentarius, EU, FAO), and was also a lecturer in food science and technology at the University of Tor Vergata (Rome, Italy) for 12 years. Since 2000, Dr. Carcea has been the Italian national delegate of the International Association for Cereal Science and Technology (ICC) in Vienna (Austria), where she has been actively involved as a member of the executive, governing, and technical committees, and was president from 2011–2012. She is also a member of the ICC Academy and of the Georgofili Academy in Florence (Italy). She is a co-founder of AISTEC, the Italian Association of Cereal Science and Technology, and served as its president from 2009 to 2015. She has been responsible for the organization of numerous successful scientific conferences worldwide, and is a member of the editorial boards of several international scientific journals. Dr. Carcea has been awarded the Harald Perten Prize and the Friedrich Scweitzer Medal for her scientific achievements.

foods

MDPI

Editorial

Nutritional Value of Grain-Based Foods

Marina Carcea

Research Centre for Food and Nutrition, Council for Agricultural Research and Economics (CREA), Via Ardeatina 546, 00178 Rome, Italy; marina.carcea@crea.gov.it; Tel.: +39-06-51494638

Received: 7 April 2020; Accepted: 10 April 2020; Published: 16 April 2020

Abstract: Grains are fundamental in the daily diets of many people worldwide; they are used for the production of popular foods, such as bread, bakery products, breakfast cereals, pasta, couscous, bulgur, and snacks. Botanically, they are the seeds of plants, belonging mainly to the groups of cereals, pseudocereals, and legumes. They contribute macronutrients to the human diet, mainly carbohydrates, but also proteins and lipids, and micronutrients, such as vitamins and minerals. They are also an important source of dietary fibre and bioactives, particularly wholegrains, which are of interest for the manufacturing of high value foods with enhanced health benefits. They can be used for the production of gluten-containing (as well as gluten-free) products. One of the main objectives of the food industry when producing grain-based foods is to manufacture safe, attractive products, with enhanced nutritional value to respond to consumer expectations. The following Special Issue "Nutritional Value of Grain-Based Foods" consists of one review and eight original research papers that contribute to the existing knowledge of important ingredients, such as fat substitutes, and of the technological quality and nutritional role of grains and grain-based foods (gluten-containing and gluten-free), such as bread, muffins, and muesli bars.

Keywords: cereals; legumes; pseudocereals; gluten-free grains; macronutrients; micronutrients; bioactives; processing; nutrition

---

Grains are the basis of daily diets for many populations worldwide. They are the seeds of plants, mainly belonging to the botanical groups of cereals, pseudocereals, and legumes.

They contribute macronutrients to the human diet, mainly carbohydrates, but also proteins and lipids, and micronutrients, such as vitamins and minerals. They are also an important source of dietary fibre and bioactives, particularly wholegrains, which are of interest for the production of high value food products with enhanced health benefits [1,2]. Many nutritional guidelines now, in several countries, recommend the inclusion of a greater proportion of wholegrains in the diet for promoting health [3–5] One of wholegrains roles, recently discovered, refers to their prebiotic activity for gut microbiota, which is fundamental for the host's well-being [6,7]. The content of the aforesaid components varies in grains, depending on genetics and growing conditions, including environment and husbandry.

Humans cannot consume grains in its raw state, so it undergoes a number of processing steps, which might include decortication, dehulling, milling, dough making, extrusion, bread making, couscous making, pasta making, noodle making, bulgur making, etc., up to home cooking [1]. Some grains, thanks to their protein composition, are suitable for the production of gluten-free foods, which are essentially eaten by people suffering from gluten intolerance [8]. Moreover, different kinds of grains can be combined in the same product to take advantage of, in some cases, the complementary composition; thus, producing food with improved nutritional value (see the combination of cereals and legumes that give origin to an excellent aminoacidic composition) [9].

The aim of this special issue was to collect studies on the latest developments in grain science, with regards, in particular, to the improvement of the nutritional value of the raw material due to breeding and/or growing conditions, and the role of processing in keeping or enhancing grains' nutritional potential for the development of healthy, attractive, and improved products (traditional or new) for human consumption.

The contribution of nine papers in this Special Issue, by 12 research groups, from institutions located in six countries, covers a number of topics connected to the nutritional value of grain-based foods, a very important area in food science. Baked food products, bread and muffins in particular, are the object of research in five papers, whereas gluten-free grains/products are covered by two papers; muesli bars and durum wheat grains are also covered by two articles.

Fat provides important sensory properties, such as colour, taste, texture, and odour to baked food products, which often contain high amounts of fat. There is growing demand by consumers for healthier products with reduced fat content, and manufacturers worldwide have started exploring substitution of fats with so-called fat replacers, which range from complex carbohydrates, gums and gels, whole food matrices, and combinations, thereof. The review by Kathryn Colla, Andrew Costanzo, and Shirani Gamlath summarizes the literature on the effect of fat replacers on the quality of baked food products [10]. The ideal fat replacers for different types of low-fat baked products were a combination of polydextrose and guar gum in biscuits at 70% fat replacement, oleogels in cake at 100% fat replacement, and inulin in crackers at 75% fat replacement. The use of oatrim (100% fat replacement), bean puree (75% fat replacement), or green pea puree (75% fat replacement) in biscuits were equally successful.

Excess sodium intake in the diet is associated with high blood pressure and risk of cardiovascular diseases. Bread has been identified as a major contributor to salt intake in the Italian diet; therefore, the research article by Marina Carcea, Valentina Narducci, Valeria Turfani, and Altero Aguzzi presents a survey of sodium chloride (common salt) content in Italian artisanal and industrial bread, to establish a baseline for salt reduction initiatives [11]. Most of the bread consumed in Italy comes from artisanal bakeries; thus, 135 samples of artisanal bread were sampled in 56 locations from Northern to Southern Italy, together with 19 samples of industrial bread representative of the entire Italian production. Salt content between 0.7% and 2.3% g/100 g (as it is basis) was found, with a mean value of 1.5%, Standard Deviation (SD) 0.3. However, the majority of samples (58%) had a content below 1.5%, with 12% having very low salt content (between 0.5 and 1.0%), whereas the remaining 42% had a salt content higher than the mean value, with a very high salt content (>2.0%) recorded for 3% of samples. With regards to industrial bread, an average content of 1.6% was found, SD 0.3. In this group, most of the samples (56%) had a very high content between 2.0 and 2.5%, whereas 5% only had a content between 1.1 and 1.5%.

Bread is also a very versatile product, which, by adequately changing ingredients, can be tailored to cater for the specific needs of some sectors of the population (e.g., the ageing). The research article by Marina Carcea, Valeria Turfani, Valentina Narducci, Alessandra Durazzo, Alberto Finamore, Marianna Roselli, and Rita Rami explores the effects of functional wheat–lentil bread on the immune functions of aged mice [12]. Legumes are considered excellent ingredients to complement cereal composition, so a functional bread, tailored for the needs of the ageing population, was baked by substituting 24% of wheat flour with red lentil flour, and compared with wheat bread. Its nutritional profile was assessed by analysing proteins, amino acids, lipids, soluble and insoluble dietary fibre, resistant starch, total polyphenols, lignans, and antioxidant capacity (Ferric Reducing Antioxidant Power assay). The wheat–lentil bread had 30% more proteins than wheat bread, a more balanced amino acids composition, almost double the minerals as well as total dietary fibre content, double the amount of polyphenols, higher amounts and varieties of lignans, and more than double the antioxidant capacity. The in vivo effect of 60-day bread consumption on the immune response was studied by means of a murine model of elderly mice. Serum cytokines and intraepithelial lymphocyte immunophenotype from the mouse intestines were analysed as markers of systemic and intestinal inflammatory status, respectively. Analysis of immune parameters in intraepithelial lymphocytes showed significant differences between the two types of bread, indicating a positive effect of the wheat–lentil bread on the intestinal immune system, whereas both breads induced a reduction in serum Interleukin-10.

Bread can also be prepared with gluten-free ingredients, such as corn starch and potato starch. The research group by Przemysław Łukasz Kowalczewski, Katarzyna Walkowiak, Łukasz Masewicz,

Olga Bartczak, Jacek Lewandowicz, Piotr Kubiak, and Hanna Maria Baranowska experimented on the substitution of starch with cricket powder as a good source of protein, fat, fibre, and minerals in gluten-free bread [13]. Levels of starch substitutions were 2%, 6%, and 10%; changes caused in the dough rheology and bread texture were studied. While the introduction of cricket powder did not greatly affect dough, the bread was instead characterised by significantly increased hardness and improved consistency. Analyses of water behaviour at the molecular level indicated that cricket powder altered both the bound and bulk water fractions. Moreover, examination of water activity revealed a decreased rate of water transport in samples of bread that contained the cricket powder.

Muffins are also popular bakery products. Generally, they contain high amounts of sugar, and their over-consumption could lead to increased health risks. For this reason, the research group of Jingrong Gao, Xinbo Guo, Margaret A. Brennan, Susan L. Mason, Xin-An Zeng, and Charles S. Brennan studied the potential of modulating reduced sugar (and the potential glycaemic response) in muffins using a combination of Stevia sweetener and cocoa powder [14]. Results illustrate that muffins with 50% replacement of sucrose were similar to the control samples in terms of volume, density, and texture. However, replacement of sugar with 100% Stevia sweetener resulted in reductions in the muffin's height, volume, and increased firmness (by four-fold) compared to the control sample. Sugar replacement significantly reduced the in vitro predictive glycaemic response of muffins (by up to 55% of the control sample).

Grains, together with a variety of other ingredients, such as fruits, nuts, seeds, and chocolate, are also used for the production of so-called muesli bars, generally consumed as snacks. In dietary guidelines across the world, they are often classified as discretionary food due to their (often) high content of fat and added sugars. A comprehensive nutrition review of grain-based muesli bars in Australia, by means of an audit of supermarket products, is provided by the research article by Felicity Curtain and Sara Grafenauer [15]. Their study aimed to provide a nutritional overview of grain-based muesli bars, comparing data from 2019 with those from 2015. Audits of grain-based Muesli bars were conducted in four major supermarkets in metropolitan Sydney, making up more than 80% of total Australian market share. Mean and standard deviation was calculated for all nutrients on-pack, including whole grain per serve and per 100 g. Compared to 2015, mean sugars declined and 31% more bars were wholegrain. Although categorized as discretionary, there were significant nutrient differences across grain-based muesli bars.

Varieties of gluten-free grains are attracting attention as raw materials to improve the nutritional quality of gluten-free foods and to relieve the monotony of a gluten-free diet. In this regard, the research group of Serena Niro, Annacristina D'Agostino, Alessandra Fratianni, Luciano Cinquanta, and Gianfranco Panfili contributed a research article on gluten-free alternative grains: nutritional evaluation and bioactive compounds [16]. The content of thiamine and riboflavin (water- soluble vitamins) as well as that of carotenoids and tocols (liposoluble vitamins) was determined on nine species of cereals and pseudocereals. The analysed samples showed a high content of bioactive compounds: in particular, amaranth, canihua, and quinoa are good sources of vitamin E, while millet, sorghum, and teff are good sources of thiamine. Moreover, millet provides a fair amount of carotenoids, in particular of lutein.

Data about the nutritional composition of gluten-free products are still limited. For this reason, Idoia Larretxi, Itziar Txurruka, Virginia Navarro, Arrate Lasa, María Ángeles Bustamante, María del Pilar Fernández- Gil, Edurne Simón, and Jonatan Miranda determined the composition of gluten-free breakfast cereals, breads, and pasta. They compared the data with equivalent gluten-containing products and were able to produce a research article on micronutrient analysis of gluten-free products. Their low content was not involved in gluten-free diet imbalance in a cohort of celiac children and adolescents [17]. Micronutrient analytical content differences (minerals and vitamins) were observed in gluten-free products when compared with their gluten-containing counterparts. In order to clarify the potential contribution of the gluten-free products to the gluten-free diet's micronutrient shortages, analytical data were used to evaluate gluten-free diets in a cohort of celiac children and adolescents. It does not seem that the lower micronutrient content of the analysed gluten-free products contributed to the micronutrient deficits detected in the gluten-free diets in this cohort (whose diets were not balanced). Nevertheless, gluten-free products (fortified for folate and biotin) are proposed to prevent the observed deficiencies.

Durum wheat is the raw material of choice for the production of popular foods worldwide, such as pasta, bread, couscous, and bulgur. With the idea of helping officials set proper quality standards for wholegrain durum wheat flours and products where the germ should be preserved, Valentina Narducci, Enrico Finotti, Vincenzo Galli, and Marina Carcea performed analyses and reported in a research article on lipids and fatty acids in Italian durum wheat (*Triticum durum* Desf.) cultivars [18]. The lipids in the durum wheat grain are, in fact, mainly present in the germ. Samples belonging to 10 popular durum wheat cultivars commercially grown in Italy were harvested and analysed for two consecutive years to account for differences due to changes in climatic conditions. Total lipid content ranged from 2.97% to 3.54% dry basis (d.b.) in the year 2010 and from 3.10% to 3.50% d.b. in the year 2011; the average value was 3.22% d.b., considering both years together. Six main fatty acids were detected in all samples in order of decreasing amounts: linoleic (C18:2) > palmitic (C16:0) ≈ oleic (C18:1) > linolenic (C18:3) > stearic (C18:0) > palmitoleic (C16:1). Significant variations in the levels of single acids between two years were observed for three samples.

The above-mentioned nine papers are the result of a variety of original researches performed worldwide on the general topic of grain science; they provide a valuable overview of current issues, which have attracted attention by the scientific community. They represent state-of-the-art research, provide us with updated knowledge, and give us useful indications on the direction of future research on grain science and technology. For these reasons, this special issue "Nutritional Value of Grain-Based Foods" is worth reading, with much attention, by experts in the field, but also by those who just want to know more about this topic.

Author Contributions: M.C. planned, wrote and reviewed this article. The author has read and agreed to the published version of the manuscript.

Funding: This Editorial received no specific funding.

Conflicts of Interest: The author declares no conflict of interest.

References

1. McKevith, B. Nutritional aspects of cereals. *Nutr. Bull.* **2004**, *29*, 111–142. [CrossRef]
2. Hall, C.; Hillen, C.; Garden Robinson, J. Composition, nutritional value and health benefits of pulses. *Cereal Chem.* **2017**, *94*, 11–31. [CrossRef]
3. EU Science Hub. Food-based Dietary Guidelines in Europe. Available online: https://ec.europa.eu/jrc/en/health-knowledge-gateway/promotion-prevention/nutrition/food-based-dietary-guidelines (accessed on 3 April 2020).
4. U.S. Department of Health and Human Services and U.S. Department of Agriculture. 2015–2020 Dietary Guidelines for Americans, 8th Edition. December 2015. Available online: https://health.gov/sites/default/files/2019-09/2015-2020_Dietary_Guidelines.pdf (accessed on 3 April 2020).
5. Dietitians Association of Australia. DAA The Australian Dietary Guidelines. Available online: https://daa.asn.au/smart-eating-for-you/smart-eating-fast-facts/healthy-eating/the-australian-dietary-guidelines/ (accessed on 3 April 2020).
6. Costabile, A.; Klinder, A.; Fava, F.; Napolitano, A.; Fogliano, V.; Leonard, C.; Gibson, G.R.; Tuohy, K.M. Whole-grain wheat breakfast cereal has a prebiotic effect on the human gut microbiota: A double-blind, placebo-controlled, crossover study. *Br. J. Nutr.* **2008**, *99*, 110–120. [CrossRef] [PubMed]
7. Harris, S.; Monteagudo-Mera, A.; Kosic, O.; Charalampopoulos, D.; Shewry, P.; Lovegrove, A. Comparative prebiotic activity of mixtures of cereal grain polysaccharides. *AMB Expr.* **2019**, *9*, 203–209. [CrossRef] [PubMed]
8. Celiac Disease Foundation. Gluten-free Foods. Available online: https://celiac.org/gluten-free-living/gluten-free-foods/ (accessed on 3 April 2020).
9. Turfani, V.; Narducci, V.; Durazzo, A.; Galli, V.; Carcea, M. Technological, nutritional and functional properties of wheat bread enriched with lentil or carob flours. *LWT-Food Sci. Technol.* **2017**, *78*, 361–366. [CrossRef]

10. Colla, K.; Costanzo, A.; Gamlath, S. Fat replacers in baked food products. *Foods* **2018**, *7*, 192. [CrossRef] [PubMed]
11. Carcea, M.; Narducci, V.; Turfani, V.; Aguzzi, A. A survey of sodium chloride content in Italian artisanal and industrial bread. *Foods* **2018**, *7*, 181. [CrossRef] [PubMed]
12. Carcea, M.; Turfani, V.; Narducci, V.; Durazzo, A.; Finamore, A.; Roselli, M.; Rami, R. Bread for the aging population: The effect of a functional wheat-lentil bread on the immune function of aged mice. *Foods* **2019**, *8*, 510. [CrossRef] [PubMed]
13. Kowalczewski, P.L.; Walkowiak, K.; Masewicz, L.; Bartczak, O.; Lewandowicz, J.; Kubiak, P.; Baranowska, H.M. Gluten-free bread with cricket powder-mechanical properties and molecular water dynamics in dough and ready product. *Foods* **2019**, *8*, 240. [CrossRef] [PubMed]
14. Gao, J.; Guo, X.; Brennan, M.A.; Mason, S.L.; Zeng, X.A.; Brennan, C.S. The potential of modulating the reducing sugar released (and the potential glycemic response) of muffins using a combination of a Stevia sweetener and cocoa powder. *Foods* **2019**, *8*, 644. [CrossRef] [PubMed]
15. Curtain, F.; Grafenauer, S. Comprehensive nutrition review of grain-based muesli bars in Australia: An audit of supermarket products. *Foods* **2019**, *8*, 370. [CrossRef] [PubMed]
16. Niro, S.; D'Agostino, A.; Fratianni, A.; Cinquanta, L.; Panfili, G. Gluten-free alternative grains: Nutritional evaluation and bioactive compounds. *Foods* **2019**, *8*, 208. [CrossRef] [PubMed]
17. Larretxi, I.; Txurruka, I.; Navarro, V.; Lasa, A.; Bustamante, M.A.; Del Pilar Fernández- Gil, M.; Simón, E.; Miranda, J. Micronutrient analysis of gluten-free products: Their low content is not involved in gluten-free diet imbalance in a cohort of celiac children and adolescent. *Foods* **2019**, *8*, 321. [CrossRef] [PubMed]
18. Narducci, V.; Finotti, E.; Galli, V.; Carcea, M. Lipids and fatty acids in Italian durum wheat (*Triticum durum* Desf.) cultivars. *Foods* **2019**, *8*, 223. [CrossRef] [PubMed]

© 2020 by the author. Licensee MDPI, Basel, Switzerland. This article is an open access article distributed under the terms and conditions of the Creative Commons Attribution (CC BY) license (http://creativecommons.org/licenses/by/4.0/).

foods

MDPI

Review

Fat Replacers in Baked Food Products

Kathryn Colla, Andrew Costanzo and Shirani Gamlath *

Centre for Advanced Sensory Sciences, School of Exercise and Nutrition Sciences, Deakin University, 1 Gheringhap Street, Geelong 3220, Australia; k.colla@deakin.edu.au (K.C.); andrew.costanzo@deakin.edu.au (A.C.)

* Correspondence: shirani.gamlath@deakin.edu.au; Tel.: +61-3-925-17267

Received: 15 October 2018; Accepted: 22 November 2018; Published: 25 November 2018

Abstract: Fat provides important sensory properties to baked food products, such as colour, taste, texture and odour, all of which contribute to overall consumer acceptance. Baked food products, such as crackers, cakes and biscuits, typically contain high amounts of fat. However, there is increasing demand for healthy snack foods with reduced fat content. In order to maintain consumer acceptance whilst simultaneously reducing the total fat content, fat replacers have been employed. There are a number of fat replacers that have been investigated in baked food products, ranging from complex carbohydrates, gums and gels, whole food matrices, and combinations thereof. Fat replacers each have different properties that affect the quality of a food product. In this review, we summarise the literature on the effect of fat replacers on the quality of baked food products. The ideal fat replacers for different types of low-fat baked products were a combination of polydextrose and guar gum in biscuits at 70% fat replacement (FR), oleogels in cake at 100% FR, and inulin in crackers at 75% FR. The use of oatrim (100% FR), bean puree (75% FR) or green pea puree (75% FR) as fat replacers in biscuits were equally successful.

Keywords: fat replacers; baked products; carbohydrates; gums; gels; whole foods

1. Introduction

Dietary fat has an important role within food matrices beyond basic nutrition. It contributes to many sensory and quality properties of a food including physical, textural and olfactory factors which all influence overall palatability. Many snack foods, in particular, rely on dietary fat to fulfil these palatable qualities in order to maintain consumer acceptance and consumption. The World Health Organisation [1], along with many national health authorities [2–5], recommends decreasing consumption of discretionary snack foods due to their poor nutritional content. Excess dietary fat intake, notably from discretionary snack foods, is one of the key contributors to excess energy intake and therefore weight gain [6]. Prevalence of overweight and obesity is rising worldwide [7,8] which is cause for concern as obesity is associated with increased risk of cardiovascular disease [9], type 2 diabetes mellitus [10], and some cancers [11].

Despite consumer awareness and product labelling [12,13], consumption of snack foods is relatively high with little compensation for the increased energy intake [14–16]. Many promoters have been attributed to increased snack food intake, such as convenience, taste, marketing and pricing [16,17]. In order to respond to these recommendations and consumer demands, manufacturing companies are increasingly developing snacks which are more nutrient dense than traditional snacks such as chips and cakes, which are typically high in added fat, sugar and sodium. Some examples of these types of innovative snacks include yoghurts, bars, puddings, crackers and chips which contain popular health foods (or superfoods) such as seeds, nuts, ancient grains, other wholegrains, dietary fibres, legumes, fruits and vegetables. While many of these snacks may be high in protein and dietary fibre, many also typically contribute large amounts of fat, sugar and sodium to the consumers' diet [18].

Efforts must be made to develop appealing snacks which are both high in protein and dietary fibre while not contributing large amounts of sodium, sugar and fat. Snack food categories such as cakes and muffins are yet to see significant innovation in creating high protein or high fibre alternatives [16]. In addition, there are still limited reduced fat options of these baked products on the market, likely attributed to the technological difficulty in producing such products. Ultimately, there is a need to increase the number of nutritious snack options available that satisfy the above drivers, while reducing fat composition and therefore total energy intake. Baked snack foods that omit dietary fat as a "low-fat" alternative often have poor sensory properties, such as crumbliness, dryness, poor mouthfeel and overall reduced consumer acceptance [19–23]. A number of potential "fat replacers" have been purported in order to reduce the fat content in food matrices whilst maintaining the sensory properties that are usually attributed to dietary fat. Fat replacers are subcategorised as either fat substitutes or fat mimetics. Fat substitutes replicate the functional and sensory properties of fat in a food, usually contain no energy or less energy than fat, and may be used to replace some or all of the fat normally present in a product [24,25]. Fat mimetics are protein- or carbohydrate-based ingredients that are not used to fully substitute the use of fat, but rather replicate some of the properties that fat provides within a food [24,25]. Many baked products on the market currently utilise fat replacers in order to reduce the total energy or fat content whilst maintaining consumer acceptance. This review aims to summarise the current evidence for application of fat replacers in biscuits, crackers, muffins, cakes and bread, and their effect on quality and sensory properties.

2. Application of Fat Replacers in Baked Products

Fat replacers are defined by the American Dietetic Association as "an ingredient that can be used to provide some or all of the functions of fat, yielding fewer calories than fat" [24]. A wide range of products in the food industry uses fat replacers, some of which include meat, dairy and baked products [24]. It is important for product developers and food technologists to understand how different fat replacers influence the sensory and physical quality of snacks in order to guide the development of healthier alternative products. For example, in cakes, fat can contribute to increased leavening, tenderness and a finer crumb through a combined effect of trapping air cells during the creaming process [26]. This structure is then set during baking due to starch gelatinisation and coagulation of egg proteins [26]. Fat is typically used in biscuits to lubricate and coat the flour granules to prevent water absorption, and the development of starch and gluten in order to achieve a fine crumb (crumbly texture) and soft, tender mouthfeel [27]. Fat also contributes other important functions to cakes, biscuits and crackers such as flavour delivery and shelf life which is achieved through delaying water absorption by starch granules [28–31].

Fat replacers can be ingredients which are of carbohydrate, protein or fat origin, with many different types of fat replacers with different structures and functions within each group. We have not differentiated fat substitutes and fat mimetics in this review as the majority of fat replacers used in baked food products are fat mimetics. Instead, we have categorised fat replacers in this review as complex carbohydrate powders, gums and gels, whole food purees and products, or a combination thereof. This categorisation is based on their functional and industrial applications rather than their chemical properties.

(a) Complex carbohydrates are typically successful fat replacers due to their ability to bind water to form a paste which can mimic the texture and viscosity of fats in food products through providing lubricant or flow properties similar to fat in some food systems [28,32]. Examples of carbohydrate based fat replacers include inulin, maltodextrin and plant fibres.

(b) Gums and gels work similarly in function to complex carbohydrates, in that they bind with water to form gels which mimic the texture and viscosity of fats [28]. While some gums and gels are made up of complex carbohydrates, this is not specific as there are some protein- and fat-based gums and gels. Examples of gums and gels used as fat replacers include pectins, oleogels and whey protein.

(c) Whole foods are complete or partial food matrices that are included in a food product as fat replacers. Recently, many products have utilised whole foods such as fruits and vegetables, legumes or cereal based ingredients as fat replacers. These foods are typically successful due to their highly creamy texture when mashed or processed. Foods such as avocado can achieve this due to its oil composition, banana for its high starch content, and legumes for their high starch and protein contents.

(d) Combinations of the above fat replacers are useful as they can potentially replicate multiple sensory qualities of dietary fat. In addition, complexes formed from these combinations, such as emulsions and esters, may have a greater fat replacing effect than the sum of their parts.

3. Summary of the Current Fat Replacers Used in Baked Products

Complex carbohydrate fat replacers range from digestible starches to non-digestible plant fibres (Table 1). It should be noted that the replacement of dietary fat with complex carbohydrates reduced energy density of all the food products in Table 1, regardless of fibre status, due to complex carbohydrate being less energy dense than fat. The use of fibres instead of starches could have an advantage on the market, as foods may meet criteria for fibre content claims. Inulin, a non-digestible dietary fibre typically derived from chicory root, was observed to have the greatest success in replacing dietary fat in baked products, where a fat replacement (FR) level of up to 75% in legume crackers and cake (1:1, inulin: water; and 1:2 inulin: water, respectively) was able to reduce total energy without any changes in consumer acceptance [33,34]. It should be noted that the addition of inulin did change the textural and physical properties of the cracker and cake products. While acceptance was not measured for the use of inulin in muffins, 50% FR had the least sensory and physical changes compared to 75% and 100% FR [35]. In addition, Zahn et al. tested the use of four commercial inulin formulations in muffins which varied in inulin to water ratio and solubility, but the outcomes for each were similar [35]. Maltodextrin was also successful at 75% fat replacer level in legume crackers and at 66% FR in muffins, although there were changes noted in aroma, appearance, taste and texture [33,36]. Total FR of inulin or maltodextrin (100%) had a significant decrease in consumer acceptance, so it is not recommended to fully replace fat in a baked food product. Results were also promising for inulin used as a fat replacer in biscuits, although there was some notable changes to textural and physical properties [33–35,37–39]. Other complex carbohydrates used as fat replacers in biscuits included lupine extract, maltodextrin, corn fibre, and rice starch, although all of these had significant effects on sensory properties of the biscuits except rice starch [33,36,40–43]. Rice starch has no significant effects on sensory properties, but was only tested at 20% FR. All complex carbohydrate fat replacers had a significant effect on the physical properties of doughs and their baked products, with significant increases in density, toughness, breaking strength, moisture, and decreases in volume for nearly all tested products [33–46].

Of all the complex carbohydrate fat replacers, inulin had the greatest success at reducing total fat and energy of the food product, with the least impact on sensory qualities and consumer acceptance, particularly in the legume crackers. Long chain inulin has the ability to form microcrystals which in turn aggregate together, interact with water, and eventually agglomerate creating a gel network [47]. To some extent, this gel network seems to have the ability to mimic the functions of fat in baked products such as being able to lubricate dry ingredients (through surrounding starch and protein), assisting in maintaining a shortening effect. Maltodextrin was also a successful fat replacer in legume crackers, although it was not as successful at replacing fat in biscuits compared to inulin. Inulin is also a good source of fibre, has promising gut health properties due to its prebiotic nature, and may increase absorption of nutrients such as calcium [48]. Moreover, inulin may benefit from marketing with fibre content claims, which may be appealing to consumers. Therefore, we recommend inulin as a reasonably high level fat replacer in crackers, cakes, biscuits and muffins [48].

Table 1. Summary of quality changes of complex carbohydrate fat replacers in baked food products.

Fat Replacer	Food(s)	FR Tested	Quality Changes
Inulin	Legume Cracker [33] Cake [34] Biscuit [37–39] Muffin [35]	25–100% 9.3–50% 35–100% 50–100%	Physical: ↓ cell density, aW, volume; ↑ breaking strength, dough consistency, moisture, crumb density, firmness, springiness; NSC cohesiveness Sensory: ↓ buttery flavour, crumbliness, acceptance (100% FR), embrowning (muffin: 50%), open surface, arched shape, typical smell, sweetness, typical taste; ↑ toasted flavour, chewiness, adhesiveness, springiness (100% FR), crispiness, crunchiness, dryness, toughness hardness, glossiness; NSC acceptance (50–75% FR)
Lupine Extract	Biscuit [40]	30–40%	Physical: ↓ volume, lightness; ↑ breaking strength, dough consistency, aW, moisture Sensory: ↓ sweetness; ↑ firmness, dryness, chewing time, roasted flavour
Maltodextrin	Biscuit [40,41] Legume Cracker [33] Muffin [36] Croissant [42]	30–40% 25–100% 66% 25–100%	Physical: ↓ volume, spread ratio; ↑ breaking strength, dough consistency, aW, moisture, cohesiveness, chewiness; NSC hardness, springiness Sensory: ↓ sweetness, overall flavour, aroma, colour, appearance, texture, taste, flavor, overall acceptance (muffin); ↑ firmness, dryness, chewing time; NSC acceptance (legumes cracker: 50–75% FR), mouthfeel
Corn Fibre	Biscuit [40]	30–40%	Physical: ↓ volume; ↑ breaking strength, dough consistency, aW, moisture Sensory: ↓ sweetness; ↑ firmness, dryness, chewing time
Rice Starch	Biscuit [43] Muffin [43]	20% 20%	Physical: ↓ volume, height (muffins); ↑ thickness (biscuit) Sensory: NSC all sensory qualities
Resistant Starch	Biscuit [34]	40%	Physical: ↓ hardness; ↑ spread ratio
Polydextrose	Biscuit [41,45,46]	11.5–50%	Physical: ↑ hardness, brittleness, aW, breaking strength; ↓ penetration distance, spread ratio [41] Sensory: ↑ hardness, ↓ overall flavour, appearance, texture, taste, acceptance, NSC colour

aW: Water activity; FR: Fat replacement; NSC: No significant change; ↓: decrease; ↑: increase.

Gum and gel fat replacers, while mostly being carbohydrates, also include lipid-based and protein-based gums and gels (Table 2). Some of these fat replacers may also increase suitability for nutrition content claims, such as sources of fibre or protein. Guar gum and xanthan gum had relatively little effect on the physical properties of the cake product when used as fat replacers [49]. While sensory measures were not compared to a control, both types of cakes were rated as acceptable with a greater acceptance in the cake containing xanthan gum, and 50% FR was considered ideal [49]. Oatrim (a tasteless white powder derived from oats, comprised of amylodextrins and 5–10% β-glucan soluble fiber; incorporated as a powder or gel), caused significant changes to the physical properties of cake, croissants and biscuits [50–52]. However, this did not appear to have any impact on the sensory properties of these foods, even at 100% FR. Pectin also caused significant changes to the physical properties of cake, croissants and biscuits, specifically increasing the hardness and reducing the volume of these foods, which was paralleled by increased perception of hardness and reduced flavour from sensory evaluations [42,46,53,54]. This is notable as pectin was tested at a relatively low FR level in cake and biscuits (10–30%), suggesting it is not an ideal fat replacer in baked products. Hydroxypropyl methylcellulose (HPMC) had significant effects on physical and sensory properties of crackers and biscuits, even at relatively low FR levels [33,55]. While consumer acceptance was not tested on crackers due to being considered unacceptable by a focus group [33], HPMC in biscuits was considered significantly less acceptable compared to control biscuits containing 18% canola oil suggesting it is also not an ideal fat replacer in these foods.

Oleogels are products of solidifying vegetable oils using natural wax esters [56–59]. The oleogelation process forms waxy crystal structure which hold liquid oil within a solid matrix, which allows the use of liquid vegetable oils in place of shortening. While this does not necessarily reduce the total fat content of a food product, it is useful in reducing saturated fat content. It should be noted that all oleogels studies reviewed in this paper did reduce overall fat content of their tested foods [56–59]. However, oleogels were not successful as fat replacers in these studies as they made biscuit and cake denser and harder. Sensory properties seemed to be promising with an increase in taste and no difference in acceptance compared to the control foods. Lastly, whey protein was also not an ideal fat replacer for biscuits as it resulted in a decrease in overall flavour and acceptance [45,46].

Overall, gums and gels were not very successful as fat replacers in baked goods. Oatrim appeared to be the most successful as there were no significant changes to the sensory properties of cake and biscuits, although there were a large range of physical changes to these foods which might have an impact on industrial applications. Xanthan gum and guar gum might potential be useful fat replacers in cake as they had little impact on physical properties, although more robust sensory evaluations are needed in future studies.

Table 2. Summary of quality changes of gum and gel fat replacers in baked food products.

Fat Replacer	Food(s)	FR Tested	Quality Changes
Xanthan Gum	Cake [49]	25–100%	Physical: ↓ volume (100% FR), elasticity; ↑ dough density; NSC aW, firmness
Guar Gum	Cake [49]	25–100%	Physical: ↓ volume (100% FR), elasticity; ↑ dough density; NSC aW, firmness
Oatrim	Cake [50] Biscuit [51,52]	20–60% 50–100%	Physical: ↓ air bubbles, viscosity, spread ratio, moisture, hardness (biscuits), brittleness; ↑ specific gravity, dough pH, height, hardness (cake), cohesiveness, springiness, aW Sensory: NSC colour, appearance, tenderness, sweetness, flavour, aftertaste, overall
Pectin	Cake [53] Biscuit [46,54] Croissant [42]	10–30% 10–100% 25–100%	Physical: ↓ spread ratio, penetration distance, volume; ↑ weight, aW, breaking strength, specific gravity, moisture, hardness Sensory: ↑ hardness, lightness, bitterness (biscuit: 100%); ↓ overall flavour, acceptance, colour, texture, cell size, taste, mouthfeel
Hydroxypropyl Methylcellulose (HPMC)	Legume Cracker [33] Biscuit [55]	25–100% 15–30%	Physical: ↓ lightness, yellowness; ↑ moisture, hardness, breaking strength; NSC aW Sensory: ↑ hardness, crispness; ↓ overall acceptance, yellowness, buttery flavour
Oleogels	Biscuit [56] Cake [57–59]	40–70% 25–100%	Physical: ↓ spread ratio, breaking strength, specific volume, fragmentation, porosity; ↑ hardness, specific gravity; NSC cell structure Sensory: ↑ hardness, chewiness, springiness, lightness (crust), colour (crust), overall taste; ↓ cohesiveness; NSC overall smell, overall acceptability
Whey Protein	Biscuit [45,46]	11.5–50%	Physical: ↓ hardness, weight; ↑ aW; NSC spread ratio Sensory: ↑ hardness; ↓ overall flavour, acceptance

aW: Water activity; FR: Fat reduction; NSC: No significant change; ↓: decrease; ↑: increase.

The interest in using whole food fat replacers has increased in recent years. These fat replacers are beneficial as they have a range of carbohydrates, lipids and proteins that may aid in the rheological properties of baked products, making them potentially more suitable than simple extracts and isolates. Overall, whole food fat replacers had the least effect on the physical and sensory properties of baked products, and in some cases increased the consumer acceptance (Table 3). Apricot kernel flour was a successful fat replacer with little impact on the physical and sensory properties of biscuits at a maximum of 50% FR [60,61]. Chia seed mucilage also had little impact on physical properties of cake and bread up to 100% FR [62,63], although sensory properties were not tested in these studies.

High oleic sunflower oil (HOSO) did not significantly decrease the amount of total fat in biscuits, but did reduce the saturated fat content [64]. However, the use of HOSO as a fat replacer was not considered successful as it has significant impact on the volume, colour and texture of the biscuits. The use of avocado puree as a fat replacer in cake and biscuits was successful at 50% FR, as it did not impact consumer acceptance [51,65]. However, at 75–100% FR, acceptance of the low-fat cake decreased compared to the control cake containing shortening [65]. Apple puree or pomace was the only whole food fat replacer to result in a reduction in sensory quality and consumer acceptance, even at low FR levels (10%) [66,67]. Therefore, apple puree is not recommended as a fat replacer in biscuits. Bean puree and green pea puree had very similar effects on the sensory properties of biscuits with increases in sensory qualities at 25–75% FR [68,69]. The use of green pea puree at FR of 25% in biscuits was considered ideal, whereas a FR of 100% resulted in reduced consumer acceptance [69]. Lastly, a high β-glucan product derived from oats or oat bran had significant impact on texture, colour and moisture of biscuits [45,54,70]. Although sensory properties were not tested in these studies, this suggests that the high β-glucan product was not a successful replacer for shortening in biscuits.

Whole foods may be the most suitable candidates for fat replacers in baked foods as they appeared to have the least impact on physical and sensory properties. In addition, they may also be beneficial as they may contain phytochemicals and micronutrients which could increase the health benefits and marketing potential of baked foods products, leading to novel functional foods. Lastly, consumer are more likely to accept foods with ingredients or additives that are made from natural, whole food products [71]. Bean and pea purees were the most successful fat replacers for biscuits at 25–75% FR, and avocado puree was successful at reducing fat in cake at 50% FR. However, more studies on whole food fat replacers in biscuits and bread is needed before they can be recommended as reliable fat replacers.

Table 3. Summary of quality changes of whole food fat replacers in baked food products.

Fat Replacer	Food(s)	FR Tested	Quality Changes
Apricot Kernel Flour	Biscuit [60,61]	10–50%	Physical: ↓ spread ratio, yellowness; ↑ hardness, lightnessSensory: NSC overall sensory score
Chia Seed Mucilage	Cake [62,63] Bread [63]	25–100% 25–100%	Physical: ↓ lightness, yellowness; ↑ firmness; NSC volume, symmetry, uniformity, redness, moisture, aW, breaking strength
High Oleic Sunflower Oil (HOSO)	Biscuit [64]	100%	Physical: ↓ volume, moisture, lightness, yellowness; ↑ biscuit density, breaking strength, redness; NSC dough density
Avocado Puree	Biscuit [51] Cake [65]	50% 50–100%	Physical: ↓ spread ratio, moisture, stiffness, hardness; ↑ aW, brittleness Sensory: ↓ appearance, acceptance (75–100%); NSC colour, tenderness, sweetness, flavour, aftertaste, acceptance (50%), overall sensory score
Apple Puree/Pomace	Biscuit [66,67]	10–100%	Physical: ↓ spread ratio, brittleness, hardness, yellowness; ↑ moisture Sensory: ↓ appearance, texture, chewiness, sweetness, moistness (100%), flavour, aftertaste, overall sensory score; ↑ moistness (50%); NSC colour
Bean Puree	Biscuit [68]	25–75%	Sensory: ↑ appearance, colour, flavour, texture, acceptance
Green Pea Puree	Biscuit [69]	25–100%	Physical: ↑ moisture Sensory: ↓ flavour, aftertaste, acceptance (100%); ↑ colour, moistness, flavour (25–75%), acceptance (25%); NSC smell
Oat Bran/High β-Glucan Oat Product	Biscuit [45,54,70]	10–100%	Physical: ↓ spread ratio, hardness [59], redness, yellowness; ↑ hardness [37], brittleness, moisture, aW, lightness, volume

aW: Water activity; FR: Fat reduction; NSC: No significant change; ↓: decrease; ↑: increase.

Fat replacers in combination with additional ingredients may provide better fat-like qualities as the additional ingredients are usually designed to supplement the unwanted effects of individual fat replacers, as seen above (Tables 1–3). These additional ingredients are usually other types of fat replacers, but can also be enzymes or emulsifiers. Few studies have assessed combined fat replacers in baked products, although the results appear promising (Table 4). Polydextrose and guar gum were successful fat replacers in biscuits at a relatively high level of FR (70%), with an increase in perceived taste, flavour and consumer acceptance [72]. Maltodextrin and xanthan gum yielded increased moisture, hardness and chewiness in 66% FR muffins, but sensory analysis was not conducted in these samples [36]. Kel-Lite BK, a commercial fat replacer containing xanthan gum, guar gum, cellulose gel, sodium stearoyl lactylate, gum Arabic, dextrin, lecithin, and mono- and diglyceride, resulted in increased bitterness and, oddly increased both crumb firmness and softness in biscuits at 33%, 66% and 100% FR [54]. HOSO and inulin were also successful fat replacers in biscuits at 100% FR [64,73], although HOSO does contain lipids so the biscuits only had reduced saturated fat rather than total fat. However, HOSO and inulin resulted in decreased appearance, flavour, odour, texture, and consumer acceptance in cakes, croissants and muffins [73]. Therefore, HOSO and inulin may only be suitable for use as fat replacers in biscuits. HOSO and β-Glucan may also be a useful fat replacer at 100% FR as this had little impact on physical properties in biscuits, although sensory evaluations were not conducted [64]. A combination of emulsion filled gel based on inulin and extra virgin olive oil (EVOO) has also been trialed as a fat mimetic in biscuits [74]. At 50% FR, there were no changes to the physical properties and the overall consumer acceptance of the biscuit compared to the control biscuit containing 20% butter, although there was a decrease in overall flavour. However, consumer acceptance was not maintained at 100% FR. Inulin, lipase and a commercial emulsifier ("Colco"; a type of alpha-gel emulsifier containing glycerol monostearate and polyglycerol esters of fatty acids) had little impact on physical properties of cake at 50–70% FR, although no sensory evaluation was conducted for this combined fat replacer either [75]. One study assessed the double, but not triple, combinations of corn fibre, maltodextrin and lupine extract in biscuits, each at 30–40% FR [40]. All combinations had little impact on the physical properties of the biscuits compared to the control biscuit containing 33% margarine. However, consumer preference for corn fibre and lupine extract was significant lower than the control, whereas corn fibre and maltodextrin was significant higher than the control [40]. This suggests that the combination of corn fibre and maltodextrin may be an ideal fat replacement in biscuits at a moderate FR level. Tapioca dextrin, tapioca starch and resistant starch as a combination fat replacer had an impact on a wide range of sensory properties in biscuits [76]. However, overall consumer acceptance decreased, even at relatively low FR levels (10–20%), so we do not recommend the use of this combination fat replacer in biscuits.

Overall, combination fat replacers may be potential candidates for ingredients in low-fat baked products. The use of polydextrose and guar gum appears to be a reasonably effective fat replacer in biscuits. However, with the limited evidence currently available, recommendations cannot be made for the use of combination fat replacers in other baked products.

Table 4. Summary of quality changes of combined fat replacers in baked food products.

Fat Replacer	Food(s)	FR Tested	Quality Changes
Polydextrose and Guar Gum	Biscuit [72]	70%	Physical: ↑ spread ratio, hardness, stress-strain ratio, moisture Sensory: ↑ overall taste, overall flavour, acceptance
Maltodextrin and Xanthan Gum	Muffin [36]	66%	Physical: ↑ aW, moisture, hardness, chewiness; ↓ volume; NSC springiness, cohesiveness
Kel-Lite BK	Biscuit [54]	33–100%	Physical: ↑ crumb firmness, crumb softness; NSC volume Sensory: ↑ bitterness

Table 4. *Cont.*

Fat Replacer	Food(s)	FR Tested	Quality Changes
HOSO and Inulin	Biscuit [64,73] Cake [73] Croissant [73] Muffin [73]	100%	Physical: NSC dough density, biscuit density, volume, moisture, breaking strength, lightness, colour Sensory: ↓ appearance (croissant and muffin), odour (croissant and muffin), texture (cake and croissant), flavour (cake and muffin), acceptance (cake, croissant and muffin), purchase intent (cake), preference (cake and muffin)
HOSO and β-Glucan	Biscuit [64]	100%	Physical: ↓ volume, lightness; ↑ biscuit density; NSC dough density, moisture, breaking strength, colour
EVOO and EFG based on Inulin	Biscuit [74]	50–100%	Physical: ↓ breaking strength (100%), porosity (100%); Sensory: ↓ caramel odour, buttery odour and flavour, sweetness, crunchiness, crush, dryness, acceptance (100%); ↑ consistency; NSC grain odour and flavour, saltiness
Inulin, Lipase and Emulsifier	Cake [75]	50–70%	Physical: ↓ batter density; ↑ cohesiveness; NSC volume, cell structure, hardness, chewiness, springiness
Corn Fibre, Maltodextrin and/or Lupine Extract	Biscuit [40]	30–40%	Physical: ↓ lightness, volume; ↑ breaking strength Sensory: ↓ preference (corn fibre and lupine extract); ↑ preference (corn fibre and maltodextrin)
Tapioca Dextrin, Tapioca Starch and Resistant Starch	Biscuit [76]	10–20%	Physical: ↓ spread ratio; ↑ breaking strength Sensory: ↓ buttery taste, crunchiness, hardness, colour, buttery odour, appearance, texture, taste, sweetness, acceptance; ↑ shape homogeneity, floury taste, pastiness, floury odour

aW: Water activity; FR: Fat reduction; NSC: No significant change; ↓: decrease; ↑: increase, HOSO: High Oleic Sunflower Oil; EVOO: Extra Virgin Olive Oil; EFG: Emulsion Filled Gel.

4. Industry Recommendations and Conclusions

It should be noted that there is limited literature on the use of fat replacers in low-fat baked products. Many of the reviewed fat replacers have only been assessed once, and also only in one type of food. There is a need for additional replicate studies using a variety of recipes. Also, while we have reviewed the current literature here, we cannot compare physical and sensory properties between studies. Therefore, while we can summarise which fat replacers were successful within a certain baked product, it is difficult to determine which fat replacer is best. In addition, the use of fat replacers in bread, muffins and croissants were only assessed in few studies each. Therefore, there is not enough information to make a recommendation of the best type of fat replacer for these products. Below is our recommendations for the best currently assessed fat replacers in a range of baked food products:

Biscuit—Oatrim was the most successful fat replacer in biscuits as it was able to retain most sensory properties of a traditional biscuit even at 100% FR, although there was a decrease in hardness and brittleness [51,52]. However, it should also be noted that both bean puree and green pea puree were able to increase the sensory qualities and consumer acceptance of biscuits at 75% FR with less of an impact on the physical properties compared to oatrim [68,69]. Legume purees might also have an advantage over oatrim as they may aid the marketability of food products due to potential nutrition claims such as vegetable and protein content. However, legume purees should not be used at 100% FR. Overall, we recommend the use of either oatrim or legume purees as fat replacers in biscuits.

Cake—Oleogels appeared to be the most successful fat replacer in cake, with no changes to the sensory qualities at 100% FR [57–59]. However, there were significant changes to the physical properties of cake when using oleogels at FR levels ≥50% [58] which might lead to difficulty during cake production. An alternative could be avocado puree which was only successful at 50% FR but had less of an impact on the physical properties of cake [65], or inulin which was successful up to 75% FR but had an impact on the physical and textural properties of cake [34].

Cracker—While there was only one study on the use of fat replacers in crackers [33], it assessed and compared a range of fat replacers in the one study. Inulin appeared to be the most successful fat replacer in these crackers, reaching an acceptable level of FR at 75%. The additional benefits of using inulin is that it may aid the marketability of food products due to potential high fibre claims.

Author Contributions: K.C. and A.C. collated the literature and wrote the manuscript. S.G. conceived the original idea, designed the format, reviewed the manuscript and provided critical feedback.

Funding: This research received no external funding.

Conflicts of Interest: The authors declare no conflict of interest.

References

1. World Health Organisation. *Prevention of Cardiovascular Disease*; World Health Organisation: Geneva, Switzerland, 2007.
2. Department of Health and Human Services, Department of Agriculture. *2015–2020 Dietary Guidelines for Americans*, 8th ed.; Department of Health and Human Services: Washington, DC, USA, 2015.
3. Health Canada. *Eating Well with Canada's Food Guide*; Health Canada: Ottawa, ON, Canada, 2011.
4. National Health and Medical Research Council. *Nutrient Reference Values for Australia and New Zealand Including Recommended Dietary Intakes*; National Health and Medical Research Council: Canberra, Australia, 2006.
5. Public Health England. *The Eatwell Guide*; Public Health England: London, UK, 2016.
6. Hooper, L.; Abdelhamid, A.; Bunn, D.; Brown, T.; Summerbell, C.D.; Skeaff, C.M. Effects of total fat intake on body weight. *Cochrane Database Syst. Rev.* **2015**, *8*, CD011834. [CrossRef] [PubMed]
7. Aune, D.; Sen, A.; Prasad, M.; Norat, T.; Janszky, I.; Tonstad, S.; Romundstad, P.; Vatten, L.J. BMI and all cause mortality: Systematic review and non-linear dose-response meta-analysis of 230 cohort studies with 3.74 million deaths among 30.3 million participants. *BMJ* **2016**, *353*, i2156. [CrossRef] [PubMed]
8. Ng, M.; Fleming, T.; Robinson, M.; Thomson, B.; Graetz, N.; Margono, C.; Mullany, E.C.; Biryukov, S.; Abbafati, C.; Abera, S.F.; et al. Global, regional, and national prevalence of overweight and obesity in children and adults during 1980–2013: A systematic analysis for the Global Burden of Disease Study 2013. *Lancet* **2014**, *384*, 766–781. [CrossRef]
9. Whitlock, G.; Lewington, S.; Sherliker, P.; Clarke, R.; Emberson, J.; Halsey, J.; Qizilbash, N.; Collins, R.; Peto, R. Body-mass index and cause-specific mortality in 900,000 adults: Collaborative analyses of 57 prospective studies. *Lancet* **2009**, *373*, 1083–1096. [PubMed]
10. Abdullah, A.; Peeters, A.; de Courten, M.; Stoelwinder, J. The magnitude of association between overweight and obesity and the risk of diabetes: A meta-analysis of prospective cohort studies. *Diabetes Res. Clin. Pract.* **2010**, *89*, 309–319. [CrossRef] [PubMed]
11. Renehan, A.G.; Tyson, M.; Egger, M.; Heller, R.F.; Zwahlen, M. Body-mass index and incidence of cancer: A systematic review and meta-analysis of prospective observational studies. *Lancet* **2008**, *371*, 569–578. [CrossRef]
12. Bucher, T.; Collins, C.; Rollo, M.E.; McCaffrey, T.A.; De Vlieger, N.; Van der Bend, D.; Truby, H.; Perez-Cueto, F.J. Nudging consumers towards healthier choices: A systematic review of positional influences on food choice. *Br. J. Nutr.* **2016**, *115*, 2252–2263. [CrossRef] [PubMed]
13. Krystallis, A.; Chrysochou, P. Do health claims and prior awareness influence Consumers' preferences for unhealthy foods? The case of functional Children's snacks. *Agribusiness* **2012**, *28*, 86–102. [CrossRef]
14. Pearson, N.; Biddle, S.J. Sedentary behavior and dietary intake in children, adolescents, and adults. *Am. J. Prevent. Med.* **2011**, *41*, 178–188. [CrossRef] [PubMed]
15. Piernas, C.; Popkin, B.M. Trends in snacking among US children. *Health Aff.* **2010**, *29*, 398–404. [CrossRef] [PubMed]
16. Forbes, S.L.; Kahiya, E.; Balderstone, C. Analysis of snack food purchasing and consumption behavior. *J. Food Prod. Market.* **2016**, *22*, 65–88. [CrossRef]
17. Chandon, P.; Wansink, B. Does food marketing need to make us fat? A review and solutions. *Nutr. Rev.* **2012**, *70*, 571–593. [CrossRef] [PubMed]
18. Block, G. Foods contributing to energy intake in the US: Data from NHANES III and NHANES 1999–2000. *J. Food Composit. Anal.* **2004**, *17*, 439–447. [CrossRef]
19. Frye, A.M.; Setser, C.S. Optimizing texture of reduced-calorie yellow layer cakes. *Cereal Chem.* **1992**, *69*, 338–343.

20. Drewnowski, A.; Nordensten, K.; Dwyer, J. Replacing sugar and fat in biscuits: Impact on product quality and preference. *Food Qual. Prefer.* **1998**, *9*, 13–20. [CrossRef]
21. Siro, I.; Kapolna, E.; Kapolna, B.; Lugasi, A. Functional food. Product development, marketing and consumer acceptance—A review. *Appetite* **2008**, *51*, 456–467. [CrossRef] [PubMed]
22. Tepper, B.J.; Trail, A.C. Taste or health: A study on consumer acceptance of corn chips. *Food Qual. Pref.* **1998**, *9*, 267–272. [CrossRef]
23. Hamilton, J.; Knox, B.; Hill, D.; Parr, H. Reduced fat products–consumer perceptions and preferences. *Br. Food J.* **2000**, *102*, 494–506. [CrossRef]
24. Richard, D.M. Position of the American Dietetic Association fat replacer. *J. Am. Diet. Assoc.* **1998**, *98*, 463–468.
25. Akoh, C.C. Fat replacers. *Food Technol.* **1998**, *52*, 47–53.
26. Wilderjans, E.; Luyts, A.; Brijs, K.; Delcour, J.A. Ingredient functionality in batter type cake making. *Trends Food Sci. Technol.* **2013**, *30*, 6–15. [CrossRef]
27. Maache-Rezzoug, Z.; Bouvier, J.M.; Allaf, K.; Patras, C. Effect of principal ingredients on rheological behaviour of biscuit dough and on quality of biscuits. *J. Food Eng.* **1998**, *35*, 23–42. [CrossRef]
28. Lucca, P.A.; Tepper, B.J. Fat replacers and the functionality of fat in foods. *Trends Food Sci. Technol.* **1994**, *5*, 12–19. [CrossRef]
29. Berenzon, S.; Saguy, I.S. Oxygen absorbers for extension of crackers shelf-life. *LWT-Food Sci. Technol.* **1998**, *31*, 1–5. [CrossRef]
30. Manley, D. *Manley's Technology of Biscuits, Crackers and Biscuits*; Woodhead Publishing: Cambridge, UK, 2011.
31. Nasir, M.; Butt, M.S.; Anjum, F.M.; Sharif, K.A.; Minhas, R. Effect of moisture on the shelf life of wheat flour. *Int. J. Agric. Biol.* **2003**, *5*, 458–459.
32. Voragen, A.G. Technological aspects of functional food-related carbohydrates. *Trends in Food Sci. Technol.* **1998**, *9*, 328–335. [CrossRef]
33. Colla, K.; Gamlath, S. Inulin and maltodextrin can replace fat in baked savoury legume snacks. *Int. J. Food Sci. Technol.* **2015**, *50*, 2297–2305. [CrossRef]
34. Rodríguez-García, J.; Puig, A.; Salvador, A.; Hernando, I. Optimization of a sponge cake formulation with inulin as fat replacer: Structure, physicochemical, and sensory properties. *J. Food Sci.* **2012**, *77*, C189–C197. [CrossRef] [PubMed]
35. Zahn, S.; Pepke, F.; Rohm, H. Effect of inulin as a fat replacer on texture and sensory properties of muffins. *Int. J. Food Sci. Technol.* **2010**, *45*, 2531–2537. [CrossRef]
36. Khouryieh, H.A.; Aramouni, F.M.; Herald, T.J. Physical and sensory characteristics of no-sugar-added/low-fat muffin. *J. Food Qual.* **2005**, *28*, 439–451. [CrossRef]
37. Błońska, A.; Marzec, A.; Błaszczyk, A. Instrumental Evaluation of Acoustic and Mechanical Texture Properties of Short-Dough Biscuits with Different Content of Fat and Inulin. *J. Texture Stud.* **2014**, *45*, 226–234. [CrossRef]
38. Rodríguez-García, J.; Laguna, L.; Puig, A.; Salvador, A.; Hernando, I. Effect of fat replacement by inulin on textural and structural properties of short dough biscuits. *Food Bioprocess Technol.* **2013**, *6*, 2739–2750. [CrossRef]
39. Krystyjan, M.; Gumul, D.; Ziobro, R.; Sikora, M. The effect of inulin as a fat replacement on dough and biscuit properties. *J. Food Qual.* **2015**, *38*, 305–315. [CrossRef]
40. Forker, A.; Zahn, S.; Rohm, H. A combination of fat replacers enables the production of fat-reduced shortdough biscuits with high-sensory quality. *Food Bioprocess Technol.* **2012**, *5*, 2497–2505. [CrossRef]
41. Sudha, M.L.; Srivastava, A.K.; Vetrimani, R.; Leelavathi, K. Fat replacement in soft dough biscuits: Its implications on dough rheology and biscuit quality. *J. Food Eng.* **2007**, *80*, 922–930. [CrossRef]
42. Shouk, A.A.; El-Faham, S.Y. Effect of fat replacers and hull-less barley flour on low-fat croissant quality. *Pol. J. Food Nutr. Sci.* **2005**, *14*, 287–292.
43. Lee, Y.T.; Puligundla, P. Characteristics of reduced-fat muffins and biscuits with native and modified rice starches. *Emir. J. Food Agric.* **2016**, *28*, 311–316. [CrossRef]
44. Basman, A.; Ozturk, S.; Kahraman, K.; Koksel, H. Emulsion and pasting properties of resistant starch with locust bean gum and their utilization in low fat biscuit formulations. *Int. J. Food Prop.* **2008**, *11*, 762–772. [CrossRef]
45. Zoulias, E.I.; Oreopoulou, V.; Tzia, C. Textural properties of low-fat biscuits containing carbohydrate-or protein-based fat replacers. *J. Food Eng.* **2002**, *55*, 337–342. [CrossRef]

46. Zoulias, E.I.; Oreopoulou, V.; Tzia, C. Effect of fat mimetics on physical, textural and sensory properties of biscuits. *Int. J. Food Prop.* **2000**, *3*, 385–397. [CrossRef]
47. Bayarri, S.; González-Tomás, L.U.; Hernando, I.; Lluch, M.A.; Costell, E. Texture perceived on inulin-enriched low-fat semisolid dairy desserts. Rheological and structural basis. *J. Texture Stud.* **2011**, *42*, 174–184. [CrossRef]
48. Shoaib, M.; Shehzad, A.; Omar, M.; Rakha, A.; Raza, H.; Sharif, H.R.; Shakeel, A.; Ansari, A.; Niazi, S. Inulin: Properties, health benefits and food applications. *Carbohydr. Polym.* **2016**, *147*, 444–454. [CrossRef] [PubMed]
49. Zambrano, F.; Despinoy, P.; Ormenese, R.C.; Faria, E.V. The use of guar and xanthan gums in the production of 'light'low fat cakes. *Int. J. Food Sci. Technol.* **2004**, *39*, 959–966. [CrossRef]
50. Lee, S.; Kim, S.; Inglett, G.E. Effect of shortening replacement with oatrim on the physical and rheological properties of cakes. *Cereal Chem.* **2005**, *82*, 120–124. [CrossRef]
51. Wekwete, B.; Navder, K.P. Effects of avocado fruit puree and oatrim as fat replacers on the physical, textural and sensory properties of oatmeal biscuits. *J. Food Qual.* **2008**, *31*, 131–141. [CrossRef]
52. Swanson, R.B.; Garen, L.A.; Parks, S.S. Effect of a carbohydrate-based fat substitute and emulsifying agents on reduced-fat peanut butter biscuits. *J. Food Qual.* **1999**, *22*, 19–29. [CrossRef]
53. Lim, J.; Ko, S.; Lee, S. Use of Yuja (Citrus junos) pectin as a fat replacer in baked foods. *Food Sci. Biotechnol.* **2014**, *23*, 1837–1841. [CrossRef]
54. Conforti, F.D.; Charles, S.A.; Duncan, S.E. Evaluation of a carbohydrate-based fat replacer in a fat-reduced baking powder biscuit. *J. Food Qual.* **1997**, *20*, 247–256. [CrossRef]
55. Laguna, L.; Primo-Martín, C.; Varela, P.; Salvador, A.; Sanz, T. HPMC and inulin as fat replacers in biscuits: Sensory and instrumental evaluation. *LWT-Food Sci. Technol.* **2014**, *56*, 494–501. [CrossRef]
56. Mert, B.; Demirkesen, I. Reducing saturated fat with oleogel/shortening blends in a baked product. *Food Chem.* **2016**, *199*, 809–816. [CrossRef] [PubMed]
57. Kim, J.Y.; Lim, J.; Lee, J.; Hwang, H.S.; Lee, S. Utilization of oleogels as a replacement for solid fat in aerated baked goods: Physicochemical, rheological, and tomographic characterization. *J. Food Sci.* **2017**, *82*, 445–452. [CrossRef] [PubMed]
58. Amoah, C.; Lim, J.; Jeong, S.; Lee, S. Assessing the effectiveness of wax-based sunflower oil oleogels in cakes as a shortening replacer. *LWT-Food Sci. Technol.* **2017**, *86*, 430–437.
59. Pehlivanoglu, H.; Ozulku, G.; Yildirim, R.M.; Demirci, M.; Toker, O.S.; Sagdic, O. Investigating the usage of unsaturated fatty acid-rich and low-calorie oleogels as a shortening mimetics in cake. *J. Food Process. Preserv.* **2018**, *42*, e13621. [CrossRef]
60. Seker, I.T.; Ozboy-Ozbas, O.; Gokbulut, I.; Ozturk, S.; Koksel, H. Utilization of apricot kernel flour as fat replacer in biscuits. *J. Food Process. Preserv.* **2010**, *34*, 15–26. [CrossRef]
61. Özboy-Özbaş, Ö.; Seker, I.T.; Gökbulut, I. Effects of resistant starch, apricot kernel flour, and fiber-rich fruit powders on low-fat biscuit quality. *Food Sci. Biotechnol.* **2010**, *19*, 979–986. [CrossRef]
62. Felisberto, M.H.; Wahanik, A.L.; Gomes-Ruffi, C.R.; Clerici, M.T.; Chang, Y.K.; Steel, C.J. Use of chia (*Salvia hispanica* L.) mucilage gel to reduce fat in pound cakes. *LWT-Food Sci. Technol.* **2015**, *63*, 1049–1055. [CrossRef]
63. Fernandes, S.S.; de las Mercedes Salas-Mellado, M. Addition of chia seed mucilage for reduction of fat content in bread and cakes. *Food Chem.* **2017**, *227*, 237–244. [CrossRef] [PubMed]
64. Onacik-Gür, S.; Żbikowska, A.; Jaroszewska, A. Effect of high-oleic sunflower oil and other pro-health ingredients on physical and sensory properties of biscuits. *CyTA-J. Food* **2015**, *13*, 621–628. [CrossRef]
65. Nguyen, D.; Carotenuto, M.; Khan, S.; Bhaduri, S.; Ghatak, R.; Navder, K.P. Effect of Avocado Fruit Puree as Fat Replacer on the Physical, Textural and Sensory Properties of Shortened Cakes. *J. Acad. Nutr. Diet.* **2013**, *113*, A59. [CrossRef]
66. Min, B.; Bae, I.Y.; Lee, H.G.; Yoo, S.H.; Lee, S. Utilization of pectin-enriched materials from apple pomace as a fat replacer in a model food system. *Bioresour. Technol.* **2010**, *101*, 5414–5418. [CrossRef] [PubMed]
67. Hayek, S.A.; Ibrahim, S.A. Consumer acceptability of chocolate chip biscuits using applesauce as a fat (butter) substitute. *Emir. J. Food Agric.* **2013**, 159–168.
68. Rankin, L.L.; Bingham, M. Acceptability of oatmeal chocolate chip biscuits prepared using pureed white beans as a fat ingredient substitute. *J. Acad. Nutr. Diet.* **2000**, *100*, 831–833.

69. Romanchik-Cerpovicz, J.E.; Jeffords, M.J.; Onyenwoke, A.C. College student acceptance of chocolate bar biscuits containing puree of canned green peas as a fat-ingredient substitute. *J. Culin. Sci. Technol.* **2018**. [CrossRef]
70. Lee, S.; Inglett, G.E. Rheological and physical evaluation of jet-cooked oat bran in low calorie biscuits. *Int. J. Food Sci. Technol.* **2006**, *41*, 553–559. [CrossRef]
71. Aschemann-Witzel, J.; Varela, P.; Peschel, A.O. Consumers' categorization of food ingredients: Do consumers perceive them as 'clean label' producers expect? An exploration with projective mapping. *Food Qual. Prefer.* **2019**, *71*, 117–128. [CrossRef]
72. Chugh, B.; Singh, G.; Kumbhar, B.K. Development of low-fat soft dough biscuits using carbohydrate-based fat replacers. *Int. J. Food Sci.* **2013**, *2013*, 576153. [CrossRef] [PubMed]
73. Doménech-Asensi, G.; Merola, N.; López-Fernández, A.; Ros-Berruezo, G.; Frontela-Saseta, C. Influence of the reformulation of ingredients in bakery products on healthy characteristics and acceptability of consumers. *Int. J. Food Sci. Nutr.* **2016**, *67*, 74–82. [CrossRef] [PubMed]
74. Giarnetti, M.; Paradiso, V.M.; Caponio, F.; Summo, C.; Pasqualone, A. Fat replacement in shortbread cookies using an emulsion filled gel based on inulin and extra virgin olive oil. *LWT-Food Sci. Technol.* **2015**, *63*, 339–345. [CrossRef]
75. Rodríguez-García, J.; Sahi, S.S.; Hernando, I. Functionality of lipase and emulsifiers in low-fat cakes with inulin. *LWT-Food Sci. Technol.* **2014**, *58*, 173–182. [CrossRef]
76. Laguna, L.; Varela, P.; Salvador, A.N.; Sanz, T.; Fiszman, S.M. Balancing texture and other sensory features in reduced fat short-dough biscuits. *J. Texture Stud.* **2012**, *43*, 235–245. [CrossRef]

© 2018 by the authors. Licensee MDPI, Basel, Switzerland. This article is an open access article distributed under the terms and conditions of the Creative Commons Attribution (CC BY) license (http://creativecommons.org/licenses/by/4.0/).

foods

MDPI

Article

A Survey of Sodium Chloride Content in Italian Artisanal and Industrial Bread

Marina Carcea *, Valentina Narducci, Valeria Turfani and Altero Aguzzi

Research Centre for Food and Nutrition, Council for Agricultural Research and Economics (CREA-AN), Via Ardeatina 546, 00178 Roma, Italy; valentina.narducci@crea.gov.it (V.N.); valeria.turfani@crea.gov.it (V.T.); altero.aguzzi@crea.gov.it (A.A.)

* Correspondence: marina.carcea@crea.gov.it; Tel.: +39-(0)6-51-494-638

Received: 19 October 2018; Accepted: 2 November 2018; Published: 5 November 2018

Abstract: A nationwide survey on salt content in both artisanal and industrial bread was undertaken in Italy to establish a baseline for salt reduction initiatives. Excess sodium intake in the diet is associated with high blood pressure and the risk of cardiovascular diseases. Bread has been identified as a major contributor to salt intake in the Italian diet. Most of the bread consumed in Italy comes from artisanal bakeries so 135 artisanal bread were sampled in 56 locations from Northern to Southern Italy together with 19 samples of industrial bread representative of the entire Italian production. Sodium chloride content was analysed according to the Volhardt's method. A salt content between 0.7% and 2.3% g/100 g (as is basis) was found, with a mean value of 1.5% (Standard Deviation, 0.3). However, the majority of samples (58%) had a content below 1.5%, with 12% having a very low salt content (between 0.5% and 1.0%), whereas the remaining 42% had a salt content higher than the mean value with a very high salt content (>2.0%) recorded for 3% of samples. As regards the industrial bread, an average content of 1.6% was found (SD, 0.3). In this group, most of the samples (56%) had a very high content between 2.0% and 2.5%, whereas 5% only had a content between 1.1% and 1.5%. Statistics on salt content are also reported for the different categories of bread.

Keywords: salt; sodium chloride; artisanal bread; industrial bread

---

1. Introduction

One third of global deaths are due to cardiovascular diseases, including heart attacks, strokes and related diseases (World Health Organization, 2007). High blood pressure is the major risk factor and, according to a substantial body of epidemiological and interventional studies, an excess of sodium in the diet is the primary cause of hypertension [1–5]. Salt intake is thus being increasingly monitored and evaluated worldwide. The human physiological need of sodium is rated around 130–230 mg/day by the World Health Organization (WHO), but in many industrialized countries sodium intake is actually 3600–4800 mg/day [6]. This indicates that the mean salt intake of populations is well in excess of dietary needs and far from the WHO recommendation to have a salt intake <5 g/day [6], that is, 2000 mg/day of sodium.

In the last decades, a wide range of initiatives aimed at salt reduction (DASH: Dietary Approaches to Stop Hypertension, WASH: World Action on Salt and Health, National Salt Reduction Weeks, CASH: Consensus Action on Salt and Health) have been started at the international level to sensitize people about salt consumption and salt content in some food categories, to educate the population about the dangers of salt in excess, and to translate scientific evidence into public health policies and plans for reformulation of processed foods [1,3,5,7–11]. In fact, processed foods are the main source of salt in the diet, with cereal products contributing the most of the overall intake [6,10,12,13], especially in those countries where bread is consumed daily at every meal. A recent survey highlighted an average yearly consumption per capita of 64 kg in Europe with Italy ranking third after Germany and France (57 kg) [14].

When in 2008 the European Commission (EC) launched the EU Framework for National Salt Initiatives, an interdisciplinary Working Group for reduction of salt intake (GIRCSI) was established in Italy at the Ministry of Health [3] with the main objective to device strategies to reduce salt consumption in the population. Bread was identified as one of the first processed foods to address and the first steps to be taken were to measure and monitor the sodium content of bread to promote reformulation of foods containing less salt. Other European countries have launched initiatives to reduce salt content in bread and recently news has appeared on the Internet that Portugal will set mandatory maximum salt levels in bread by 2019 [15].

This paper represents the first comprehensive survey on the salt content in bread consumed by the Italian population, and the data reported here represent the baseline for the reformulation of salt reduced bread. Most of the bread consumed in Italy is produced by artisans in artisanal bakeries according to different recipes and procedures and only a small proportion of the market (around 10%) is covered by the industrial production: consequently, a great variability in salt content was expected. Several breads in Italy are also protected by European authenticity labels such as Protected Designation of Origin (PDO) and Protected Geographical Indication (PGI) labels. Both artisanal and industrial bread was considered in the present study. Moreover, a comparison between methods to determine Na content in flour and bread was made on selected bread samples to assess the reliability of the quick method which was used for sodium chloride determination in bread.

2. Materials and Methods

2.1. Samples and Sampling Method

Artisanal bread was purchased at selected bakeries in Northern, Central and Southern Italy particularly in places with a specific identity in terms of bread production. In each bakery, the most consumed types of bread were sampled. For the industrial sector, samples of all the Italian production available on the market were purchased at supermarkets and included sliced pan bread (12 samples) and "traditional-like" bread (6 samples). In total, 154 bread samples (kinds of bread) were collected, between winter 2009 and spring 2010. For each type of sample, a spreadsheet was filled with data concerning origin, ingredients, weight and baking method.

In detail, 19 samples of industrial bread were collected together with 135 samples of artisanal bread from 56 locations (Figure 1). Seven out of 154 samples (1 sample of industrial bread and 6 samples of artisanal bread) were declared, at purchase, without salt and subsequent analysis performed by us, confirmed this feature.

Samples of baking wheat flour (*Triticum aestivum* L. flour, which is the kind of flour mostly used in bread baking in Italy) of two different extraction rates according to the Italian law (0 and 00, ash content maximum 0.65% and 0.55% on dry matter, respectively) were purchased at a local supermarket and analysed for their sodium content.

2.2. Analytical Methods

Soon after purchase, representative portions of each type of bread were cut in small pieces, well homogenised and used for the following analyses. A portion of the sample was used to determine moisture according to ICC Standard No. 110/1 [16], whereas another portion was prepared according to AACC method 62-05 "Preparation of sample: bread" [17] by drying it at 35 °C overnight and grinding it by a MLI 204 laboratory mill (Bühler, Uzwil, Switzerland). The residual moisture in the sample was also determined according to the previous ICC Standard. The determination of chloride ion in bread samples was carried out by titration according to the AACC method 40-33 "Chloride in yeast foods—quantitative method (Volhardt's method)" [17]. Sodium chloride content was finally calculated based on the content of chloride ions in sample. Duplicate analysis was carried out for each sample. Duplicates differing by more than 0.20 were rejected and analysis repeated. Salt content in bread was expressed as percentage, as is basis.

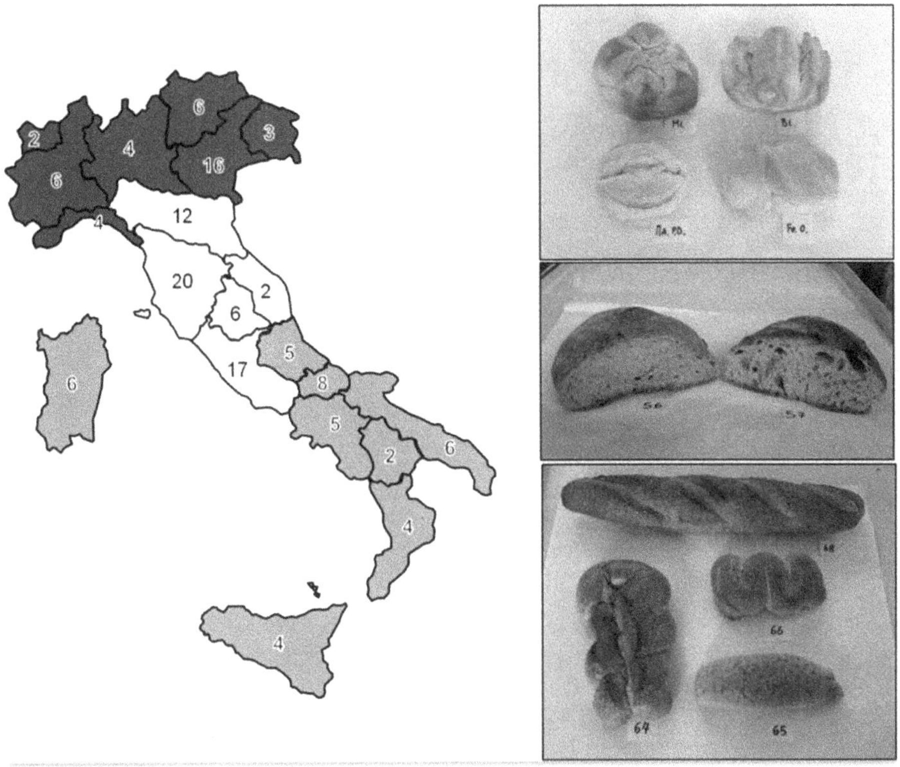

(**a**) (**b**)

Figure 1. Bread samples collected in the different regions of Northern, Central and Southern Italy. (**a**) Number of samples from each Italian region. Northern regions are coloured in dark grey, Central regions in white and Southern regions in light grey (division according to the Italian Central Institute of Statistics). (**b**) Bread samples of different size, shape and ingredients.

A selection of bread and wheat flour samples were also analysed by Inductively Coupled Plasma Spectroscopy (ICP) on a Perkin-Elmer Plasma Optima 3200XL (Perkin-Elmer, Waltham, MA, USA) in order to determine sodium content in the raw material, and confirm that the results obtained by the AACC method were in good match with those obtained by ICP. Samples were first mineralized in nitric acid (6 mL HNO_3 + 1 mL H_2O_2) in a microwave oven (Milestone 1200 Mega, FKV srl, Torre Boldone, Italy). Standard CRM 189 (whole meal flour) from the Community Bureau of Reference (BCR, Brussels, Belgium) was used as a Reference Material.

2.3. Statistics

The seven samples of bread without salt were excluded from statistical elaboration. Statistical determination of mean, standard deviation and percentage distribution were performed using Microsoft Office Excel 2007. For easiness of results understanding and interpretation, it was decided to establish 4 classes of salt content (as is basis): (i) 0.5–1.0% (low salt content); (ii) 1.1–1.5% (medium salt content); (iii) 1.6–2.0% (high salt content); and (iv) 2.1–2.5% (very high salt content).

The percentage distribution in the above-mentioned salt content classes was calculated for 14 groups that represented all the different commercial categories that could be found in our sample population: all samples together, industrial vs. artisanal samples, 4 categories according to weight, 5 categories according to ingredients, and 2 categories according to leavening method.

3. Results

This section presents the results of analyses of sodium content in soft wheat white flour widely used for bread baking in Italy, and eight samples, selected for their different characteristics and presumably different salt content, are reported in Table 1. The same table briefly describes each sample compositional or processing characteristics. One column reports data obtained by calculating the sodium content in samples analysed by the standard AACC method 40-33 (Volhardt's method) [17], whereas the other column refers to the sodium content in samples determined by means of ICP.

The purpose of this study was to assess the contribution of the raw material flour to the salt content in bread, to verify whether the bread declared to be without any salt actually had a negligible sodium content, and whether the data obtained by the Volhardt's method could be compared with those obtained by a more sensitive but more complex and expensive method.

Data reported in Table 1 show that sodium was not detected in both types of commercial soft wheat white flours, even in the 0 type which is less refined than 00. Based on this result obtained with a very sensitive instrument, it was decided not to analyse these two samples by the Volhardt's method.

No sodium was detected following both analytical procedures in the three different bread samples declared by the bakers to be without salt addition. Sodium was detected by means of both methods in the five remaining samples and values ranged 03–06 for the Volhardt's method and 0.1970–0.4902 g/100 g (as is sample) for the ICP method. In both cases, the highest value was obtained for durum wheat bread.

Table 1. Sodium content in flour and bread samples as measured by two methods of different sensitivity.

Sample	Sodium Content Volhardt * (g/100 g)	Sodium Content ICP * (g/100 g)
Commercial white flour (Italian type 00)	not analysed	not detected
Commercial white flour (Italian type 0)	not analysed	not detected
Sample 1 (bread without salt, 500 g)	not detected	not detected
Sample 4 (bread without salt, 500 g)	not detected	not detected
Sample 14 (bread without salt, 500 g)	not detected	not detected
Sample 32 (common bread, 95 g)	0.3	0.2249
Sample 38 (wholemeal sourdough bread, 1.5 kg)	0.4	0.3283
Sample 25 (durum wheat bread, 170 g)	0.6	0.4902
Sample 57 (sourdough bread, 2 kg)	0.3	0.1970
Sample 67 (special bread, 200 g)	0.6	0.4552

* Average of two determinations on as is sample.

The statistical elaboration of salt content data referring to the 147 samples of salty bread is reported in Figures 2–5. In our survey, a salt content in bread ranging between 0.7% and 2.3% (as is basis) was found, with a mean value of 1.5% and a standard deviation (SD) of 0.3 (Figure 3). If we look at the distribution of salt content in the different classes as specified in the Materials and Methods Section, we can see that the majority of bread samples (58%) had a salt content below the reported mean value (>1.5%) (Figure 2a) with 12% having a very low salt content falling within the range 0.5–1.0%, whereas the remaining 42% had a salt content higher than the mean value with a very high salt content (>2.0%) recorded for 3% of samples.

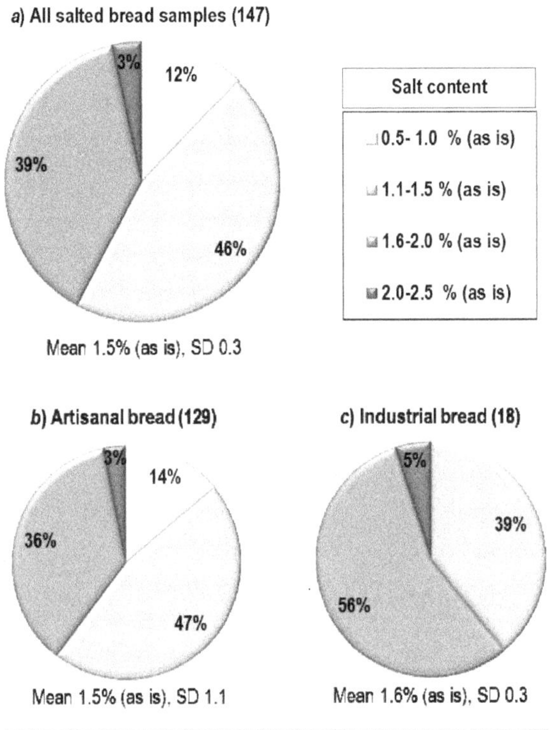

Figure 2. Percent distribution of bread samples according to salt content classes.

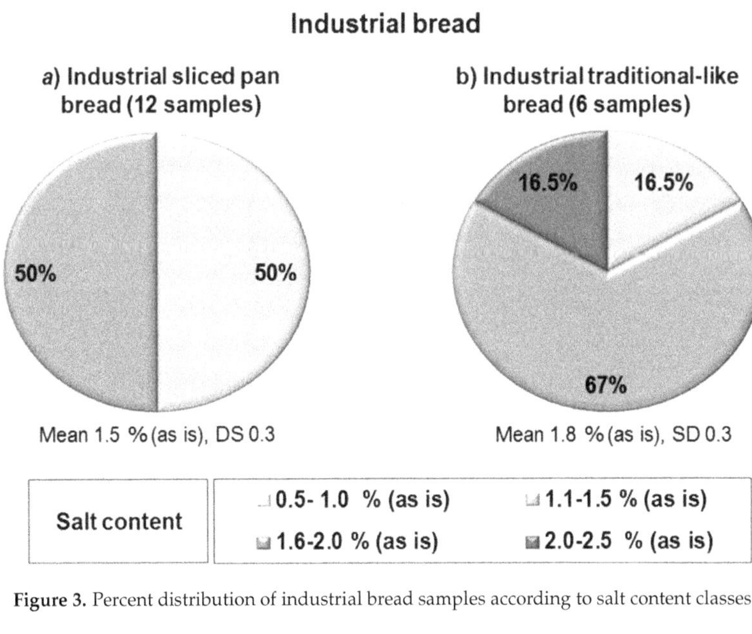

Figure 3. Percent distribution of industrial bread samples according to salt content classes.

If we have a separate look at the artisanal and the industrial production (Figure 2b,c), we can say that, although the average salt content in bread is very similar (1.5 and 1.6 g/100 g as is basis with standard deviations of 1.1 and 0.3, respectively), the distribution of our samples in the different salt content classes is different. In the artisanal bread, the majority of bread samples (61%) had a salt content below the reported mean value (>1.5%) (Figure 2a) with 14% having a very low salt content falling within the range 0.5–1.0%, whereas the remaining 39% had a salt content higher than the mean value with a very high salt content (>2.0%) recorded for only 3% of samples. In the industrial production, only three classes were represented, the very low salt content class (<1.0 g/100 g as is) having disappeared. Most of the samples (56%) had a very high salt content between 2.0 and 2.5 g/100 g (as is) whereas only 5% had a salt content between 1.1 and 1.5 g/100 g (as is).

A further differentiation can be made within the industrial bread by considering separately the sliced pan bread, which represents the most consumed category, and the so-called "traditional-like" bread which resembles more in its shape the artisanal bread (Figure 3). In the pan bread, a mean value of 1.5 g/100 g (as is) was obtained (SD 0.3) and two salt content classes (1.1–2.0%) were found, each having a 50% share, whereas in the traditional-like bread an average value of 1.8% g/100 g (as is) was found (SD 0.3) which derived from the contribution of three salt content classes (1.1–2.5%) with the very high salt content class having a share of 16.5%.

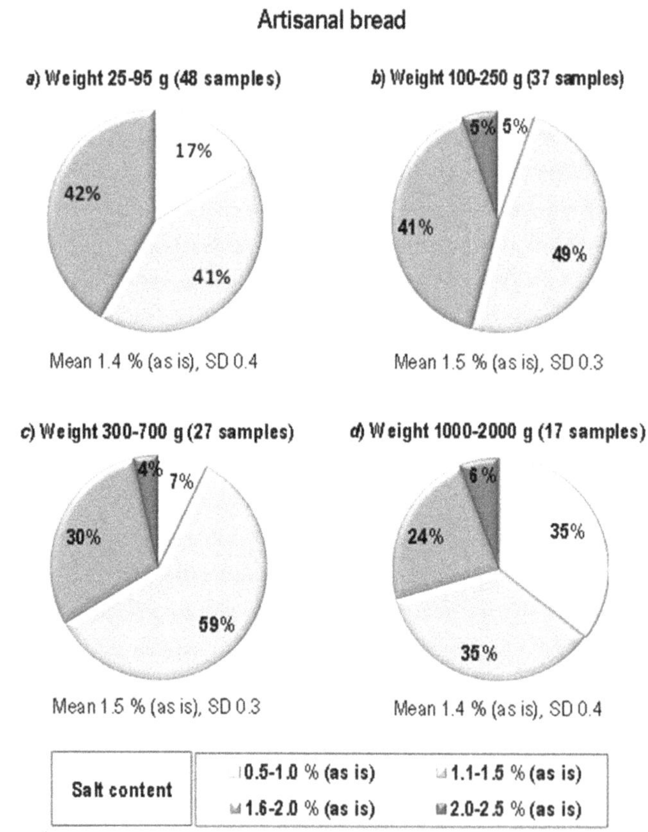

Figure 4. Percent distribution of artisanal bread samples of different weight according to salt content classes.

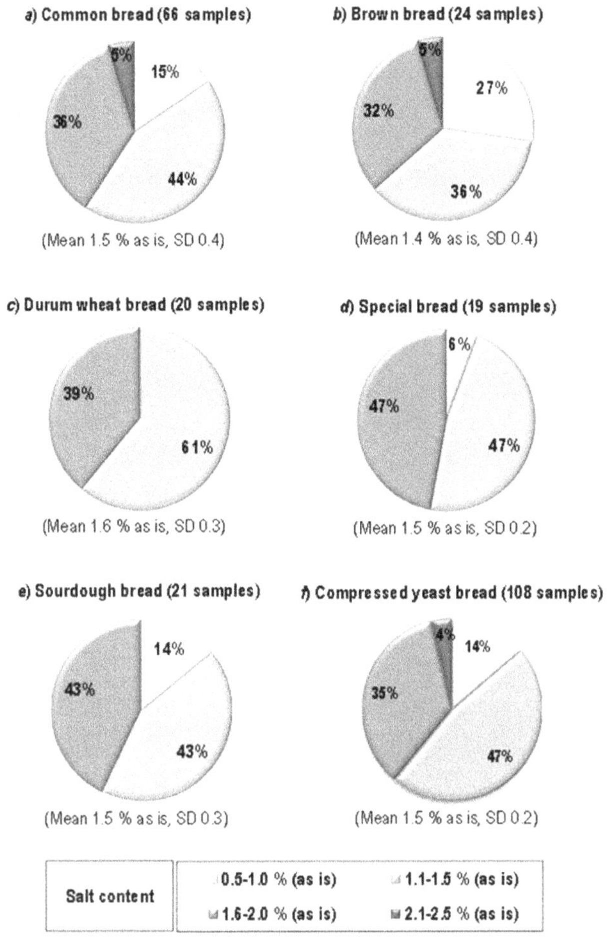

Figure 5. Per cent distribution of artisanal bread samples, differing in dough formulation and leavening method, according to salt content classes.

Given the great variety of artisanal bread, we thought it would be interesting to compare the salt content in different types of bread to determine whether there was any relationship between specific bread characteristics and salt content: weight, ingredients and leavening method were identified as interesting quality traits. The 129 artisanal bread loaves were, therefore, grouped into three different categories according to their weight, ingredients and leavening method. Within the "weight" category, four classes were identified based also on the bread shape: (i) 25–95 g (48 samples); (ii) 100–250 g (37 samples); (iii) 300–700 g (27 samples); and (iv) 1000–2000 g (17 samples), with rolls, typical of the bread production in Northern Italian regions, and big loaves typical of Central and Southern regions. Four classes were also established in the "ingredients" category as follows: common white bread, whose dough is typically formulated with just soft wheat flour, water and salt (66 samples); brown bread, with different amounts of soft wheat whole-meal flour in addition to the common white bread ingredients (24 samples); durum wheat bread (20 samples), typical of Southern Italy but also appreciated and consumed all over Italy made with remilled durum wheat semolina, water and salt; and "special" bread, that is, soft wheat white bread with other ingredients such as oil, milk, and potatoes (19 samples). As regards the leavening method, two classes were established: sourdough and compressed yeast.

Figure 4 reports the pie charts of the percentage distribution in the four salt content classes according to the weight of the bread. The most represented weight class was small breads, i.e., rolls (48 samples), and the least represented was big loaves weighing up to 2 kg. This distribution actually reflects the pattern of consumption of the Italian population. The categories up to 250 g were the most represented. Although the average salt content and SD is very similar or identical in the four groups and goes from 1.4 to 1.5 g/100 g (as is), (SD, 03 and 0.5, respectively), the percentage distribution of the four salt content classes was different and peculiar within each group with the highest salt content class not being represented for example in the smallest bread group and the biggest loaves having the highest percentage of samples (6%) having a salt content between 2.0% and 2.5% (as is).

For dough formulation (Figure 5), we obtained a mean value of 1.4 g/100 g (as is) (SD, 0.4), for brown bread, 1.5 g/100 g (as is) (SD, 0.4 and 0.2, respectively, for common bread and special bread), and for durum wheat bread, 1.6 g/100 g (as is) (SD, 0.3). In the durum wheat group, only two salt classes were found, namely 1.1–1.5% and 1.6–2.0%, with the first being more represented (61%) than the latter (39%).

The two leavening methods had very different sizes, with sourdough samples being only 21 while compressed yeast bread samples being 108. These numbers actually reflect the presence of these categories on the market with sourdough bread being less frequently found. However, the two groups had the same average salt content, 1.5 g/100 g (as is) (with SD = 0.3 for sourdough bread, and SD = 0.2 for compressed yeast bread). In the sourdough bread group, there were no samples with a very high salt content ($\geq$2.1%).

4. Discussion

Recently, several similar surveys have been conducted in countries where bread is a staple food and has therefore been identified as a major contributor to the daily intake of salt and sodium in the population [18–21].

In our study, the analysis of sodium content in a selection of commercial refined wheat flour and bread samples by ICP analysis showed that salt content in white bread, which is the most consumed type of bread in Italy, is not due to a natural occurrence of sodium in the flour, but to the salt added in the recipe. The higher sensitivity of the ICP analysis than the Volhardt's method enabled to confirm, in fact, that sodium naturally occurring in the white flour is negligible (Table 1) and, moreover, it showed that salt content in some of the sampled bread samples, declared at purchase to be "without salt", was, in fact, below 0.1%.

Even if there is no perfect correspondence between the results obtained by the two methods (Table 1), it is nevertheless interesting to notice that the ranking of the samples as regards their sodium content was the same. These results confirmed the practical value of the Volhardt's method for the determination of sodium chloride in bread and for the purpose of our study.

Although the average salt content found in all our bread samples (1.5% g/100 g, as is basis) is similar to that reported in the literature for other European countries [22], the range of values found was very wide with the highest values around 2.3%. This means that there is room for improvement and that salt reduction initiatives and campaigns are advisable also in Italy.

The statistical elaboration of data also showed an interesting variation of salt content in bread at geographical level. It emerged that the mean salt content in bread produced and consumed in Central Italy is slightly lower than in the north and south of the country. In fact, the mean salt content was 1.3% in the 52 bread samples from Central Italy with a SD of 0.4, whereas it was 1.6% with a SD of 0.2 in the 38 bread samples from Northern Italy, and 1.5% with a SD of 0.3 in the 39 bread samples from Southern Italy. In detail, it emerged that in Northern Italy there is no share of bread with a salt content below 1.0%, whereas 21% of analysed samples purchased in Central Italy and 15% of bread types sampled in Southern Italy were in this range. These figures confirm the existence of a well-established tradition in some regions of central Italy, e.g., Umbria, Marche and Tuscany, of producing bread loaves with a very low or null salt content. This evidence also hints at the fact that the main problem in salt reduction might be consumers' acceptance and salt content in bread might be reduced at the artisanal level without encountering too many technological problems.

Considering separately the artisanal production from the industrial production, even though in Italy the latter represents one fourth of the former, it is interesting to notice that the average salt content is higher in industrial bread (1.6% g/100 g, as is basis, with a SD of 0.3) than in the artisanal bread (1.5% g/100 g, as is, and a wider SD 1.1). and no samples were found falling within the class containing a small amount of salt (0.5–1.0%). Most industrial samples (56%) fall in the high salt content class (1.6–2.0%), whereas artisanal bread's most represented category (47%) is that of 1.1–1.5% salt content (medium salt content). The industrial production can easily be subdivided into two categories, namely pan bread (which is always sliced) and traditional-like bread which is more similar in shape and appearance to artisanal bread. They represent the two most common types of industrial bread that are produced by a few manufacturers in a homogeneous and standardized way, and distributed all over the national territory. It is interesting to notice that the pan bread had a more homogeneous salt content, ranging from 1.1% to 2.0% with an average of 1.5%, as is, and a SD of 0.3, whereas the traditional-like bread had 16.5% of samples having a salt content between 2.1% and 2.5% and a higher average content of 1.8% and the same value (0.3) of standard deviation.

The average content in Italian industrial bread is higher than that reported in other European countries such as UK, where in 2011 a National survey, promoted by the Consensus Action on Salt and Health (CASH), reported for industrial pre-packaged bread a salt content ranging between 0.58% and 0.83% [7].

In addition, in the industrial Italian production, it is advisable to reduce the salt content and, considering that most of the production is in the hands of few manufacturers, it should not be too difficult to reach this target. Moreover, being industrial bread generally supplied to canteens, hospitals and caterings, there are high chances that salt reduction initiatives can reach a broad number of consumers in a very short time even if the artisanal market share represents the biggest challenge for any future salt reduction initiative.

The analysis of salt content in bread according to its weight showed two significant pieces of evidence. In big loaves weighing 1000–2000 g (Figure 4d), there is a more consistent percentage of samples (35%) with a very low salt content (0.5–1.0%, as is basis). On the other hand, rolls weighing 25–95 g (Figure 4a) proved to be the only weight class with a salt content always below 2.0% and never reaching the very high content. Comparing the results obtained for the four classes under consideration with the mean salt content obtained for artisanal bread (1.5%, as is basis, with SD of 1.1), it emerged that a very good share of samples for each class has values below this mean: 57% of rolls (class 25–95 g), 54% of small loaves (class 100–250 g), 66% of medium loaves (class 300–700 g) and 70% of big loaves (class 1000–2000 g).

Considering dough formulation, i.e., the different raw materials used in bread making (Figure 5), it emerged that durum wheat bread had a more homogeneous salt content than common, brown or special bread: all samples belonged to only two salt classes, namely 1.1–1.5% and 1.6–2.0%. The main share (61%) is due to the lower salt content class. Considering the mean value of salt content in artisanal bread as a reference point for discussion, it was observed that 59% of common bread (Figure 5a), 63% of brown bread (Figure 5b), 61% of durum wheat bread (Figure 5c) and 53% of special bread (Figure 5d) samples have a salt content lower than this mean.

Bread samples with a salt content exceeding 2% belonged only to the class "common bread" and "brown bread", but at the same time brown bread is the category with the highest percentage (27%) of samples with a very low salt content (0.5–1.0%, as is basis) followed by common bread (15% of samples). The main difference between the sourdough and compressed yeast bread categories can be seen in the presence of 4% samples with a very high salt content (2.1–2.5%). The average content is the same for both categories, i.e., 1.5%, as is, but the SD is higher (0.3 versus 0.2) for sourdough bread. By focusing on the results obtained for the sourdough bread and brown bread categories, which had a significant percentage of the very low salt content, it could be speculated that the use of the sourdough and the formulation with wholemeal flours, can add to bread a natural flavour that prevents an excessive addition of salt to the dough.

5. Conclusions

The present study represents the first extensive survey on the actual salt content in Italian bread and provides the baseline for national salt reduction initiatives, as recommended by the European Commission (EC) to each country within the EU Salt Reduction Framework [8].

As regards artisanal bread, which is the type of bread mostly consumed by the Italian population, the survey highlighted a great variability of values obtained for salt content (from 0.7% to 2.3%, as is basis) that enabled both the identification of a market share offering bread with a high-salt content (2.0–2.5%) that should be immediately addressed by salt reduction policies and education campaigns, as well as the existence of a substantial share of bread with a low salt content that is in line with the EC and WHO recommendations. A good share of the Italian bakery market is represented by the long-established tradition of bread produced with a low salt content (0.5–1.0%) and widely consumed in some regions of Central Italy, e.g., Marche, Toscana and Umbria. This evidence indicates that technological strategies for low-salt bread manufacturing and campaigns for consumer education to gradual salt reduction in bread are possible with high chances of success.

As regards industrial bread, there is less variation in salt content compared to artisanal bread but it is on the high content side. However, future initiatives for salt reduction are more likely to be successful and reach in shorter times a major share of consumers because industrial bread production is controlled by a few manufacturers that distribute their standardized products all over Italy.

Author Contributions: M.C. planned and led the research and wrote the paper. M.C., V.N. and V.T. sampled the bread. V.N. and V.T. performed the experiments helped by A.A. who performed the ICP analysis. V.N. performed the statistical calculations and produced the figures.

Funding: This work was carried out within the framework of the MINISAL project funded by the Italian Ministry of Health.

Acknowledgments: The authors wish to thank Vittorio Vivanti (INRAN, Italy) and Licia Iacoviello (University of Molise, Italy) for their valuable collaboration in bread sampling, Paolo Fantauzzi for his technical assistance and Francesca Melini for her bibliographical help.

Conflicts of Interest: The authors declare no conflict of interest.

References

1. He, F.J.; MacGregor, G.A. A comprehensive review on salt and health and current experience of worldwide salt reduction programmes. *J. Human Hypertens.* **2009**, *23*, 363–384. [CrossRef] [PubMed]
2. Campbell, N.; Correa-Rotter, R.; Neal, B.; Cappuccio, F.P. New evidence relating to the health impact of reducing salt intake. *Nutr. Metab. Cardiovasc. Dis.* **2011**, *21*, 617–619. [CrossRef] [PubMed]
3. Strazzullo, P.; Cairella, G.; Campanozzi, A.; Carcea, M.; Galeone, D.; Galletti, F.; Giampaoli, S.; Iacoviello, L.; Scalfi, L. Population based strategy for dietary salt intake reduction: Italian initiatives in the European framework. *Nutr. Metab. Cardiovasc. Dis.* **2011**, *22*, 161–166. [CrossRef] [PubMed]
4. Wyness, L.A.; Butriss, J.L.; Stanner, S.A. Reducing the population's sodium intake: The UK Food Standards Agency's salt reduction programme. *Public Health Nutr.* **2011**, *15*, 254–261. [CrossRef] [PubMed]
5. Brindsen, H.C.; Farrand, C.E. Reducing salt; preventing stroke. *Nutr. Bull.* **2012**, *37*, 57–63.
6. World Health Organisation. Reducing Salt Intake in Populations. Report of a WHO Forum and Technical Meeting. Geneva, Switzerland, 2007. Available online: http://www.who.int/dietphysicalactivity/Salt_Report_VC_april07.pdf (accessed on 19 October 2018).
7. CASH Consensus Action on Salt and Health. 1996. Available online: http://www.actiononsalt.org.uk/about (accessed on 19 October 2018).

8. European Union. Implementation of the EU Salt Reduction Framework. Results of Member States Survey. 2012. Available online: http://ec.europa.eu/health/nutrition_physical_activity/high_level_group/nutrition_salt_en.htm (accessed on 19 October 2018).
9. He, F.J.; MacGregor, G.A. Reducing population salt intake worldwide: From evidence to implementation. *Prog. Cardiovasc. Dis.* **2010**, *52*, 363–382. [CrossRef] [PubMed]
10. Quilez, J.; Salas-Salvado, J. Salt in bread in Europe: potential benefits of reduction. *Nutr. Rev.* **2011**, *70*, 666–678. [CrossRef] [PubMed]
11. WASH World Action on Salt and Health. 2012. Available online: http://www.worldactiononsalt.com/ (accessed on 19 October 2018).
12. Lynch, E.J.; Dal Bello, F.; Sheehan, E.M.; Cashman, K.D.; Arendt, E.K. Fundamental studies on the reduction of salt on dough and bread characteristics. *Food Research Int.* **2009**, *42*, 885–891. [CrossRef]
13. Dötsch, M.; Busch, J.; Batenburg, M.; Liem, G.; Tareilus, E.; Mueller, R.; Meijer, G. Strategies to reduce sodium consumption: A food industry perspective. *Crit. Rev. Food Sci. Nutr.* **2010**, *49*, 841–851. [CrossRef] [PubMed]
14. Food. Pane Italia al terzo posto in Europa per i consumi. 2018. Available online: https://www.foodweb.it/2016/06/pane-italia-al-terzo-posto-per-i-consumi/ (accessed on 19 October 2018).
15. Food Navigator. Portugal to Set Mandatory Maximum Salt Levels in Bread. 2018. Available online: https://www.foodnavigator.com/Article/2018/07/23/Portugal-to-set-mandatory-maximum-salt-levels-in-bread# (accessed on 19 October 2018).
16. ICC. *Standard Methods of the International Association of Cereal Science and Technology*; The Association: Vienna, Austria, 2003.
17. AACC. *Approved methods of the American Association of Cereal Chemists*, 10th ed.; The Association: St. Paul, MN, USA, 2005.
18. Jafri, A.; El-Kardi, Y.; Derouiche, A. Sodium chloride composition of commercial white bread in Morocco. *East. Mediterr. Health J.* **2017**, *23*, 708–710. [CrossRef] [PubMed]
19. Al Jawaldeh, A.; Al-Khamaiseh, M. Assessment of salt concentration in bread commonly consumed in the Eastern Mediterranean region. *East. Mediterr. Health J.* **2018**, *24*, 18–24. [CrossRef] [PubMed]
20. Coyne, K.J.; Baldridge, A.S.; Huffman, M.D.; Jenner, K.; Xavier, D.; Dunford, E.K. Differences in the sodium content of bread products in the USA and UK: Implications for policy. *Public Health Nutr.* **2018**, *21*, 632–636. [CrossRef] [PubMed]
21. Pérez Farinós, N.; Santos nSanz, S.; Dal Re, M.-A.; Yusta Boyo, M.J.; Robledo, T.; Castrodeza, J.J.; Campos Amado, J.; Villar, C. Salt content in bread in Spain, 2014. *Nutr. Hosp.* **2018**, *35*, 650–654. [CrossRef] [PubMed]
22. Joossens, J.V.; Sasaki, S.; Kesteloot, H. Bread as a source of salt: an international comparison. *J. Am. Coll. Nutr.* **1994**, *13*, 179–183. [CrossRef] [PubMed]

© 2018 by the authors. Licensee MDPI, Basel, Switzerland. This article is an open access article distributed under the terms and conditions of the Creative Commons Attribution (CC BY) license (http://creativecommons.org/licenses/by/4.0/).

foods

MDPI

Article

Bread for the Aging Population: The Effect of a Functional Wheat–Lentil Bread on the Immune Function of Aged Mice

Marina Carcea *, Valeria Turfani, Valentina Narducci, Alessandra Durazzo, Alberto Finamore, Marianna Roselli and Rita Rami

Research Centre for Food and Nutrition, Council for Agricultural Research and Economics (CREA), via Ardeatina 546, 00178 Roma, Italy; valeria.turfani@crea.gov.it (V.T.); valentina.narducci@crea.gov.it (V.N.); alessandra.durazzo@crea.gov.it (A.D.); alberto.finamore@crea.gov.it (A.F.); marianna.roselli@crea.gov.it (M.R.); rita.rami@crea.gov.it (R.R.)

* Correspondence: marina.carcea@crea.gov.it; Tel.: +39-06-5149-4638

Received: 24 September 2019; Accepted: 16 October 2019; Published: 18 October 2019

Abstract: A functional bread tailored for the needs of the aging population was baked by substituting 24% of wheat flour with red lentil flour and compared with wheat bread. Its nutritional profile was assessed by analysing proteins, amino acids, lipids, soluble and insoluble dietary fibre, resistant starch, total polyphenols, lignans and the antioxidant capacity (FRAP assay). The wheat–lentil bread had 30% more proteins than wheat bread (8.3%, as is), a more balanced amino acids composition, an almost double mineral (0.63%, as is) as well as total dietary fibre content (4.6%, as is), double the amount of polyphenols (939.1 mg GAE/100g on dry matter, d.m.), higher amounts and variety of lignans, and more than double the antioxidant capacity (71.6 μmoL/g d.m.). The in vivo effect of 60 days bread consumption on the immune response was studied by means of a murine model of elderly mice. Serum cytokines and intraepithelial lymphocyte immunophenotype from the mice intestine were analysed as markers of systemic and intestinal inflammatory status, respectively. Analysis of immune parameters in intraepithelial lymphocytes showed significant differences among the two types of bread indicating a positive effect of the wheat–lentil bread on the intestinal immune system, whereas both breads induced a reduction in serum IL-10.

Keywords: wheat bread; lentil bread; bread composition; aged mice; immune function; intraepithelial lymphocytes; gut health

1. Introduction

According to recent statistics [1], the European population in particular is an aging one: In fact, the proportion of the population over 65 has steadily increased over the past decade. Aging is a condition that brings about a number of factors contributing to the risk of malnutrition which are related to physiological changes and medical and social conditions [2,3]. Current data for mean nutrient intakes suggest that, as a group, older adults are at risk of not meeting the recommended dietary allowance (RDA) or adequate intake (AI) values for calcium, vitamins, minerals, fibre [4] and protein [5]. It is well known that increased thresholds for taste and smell resulting in bland and uninteresting food tasting, coupled with impaired masticatory efficiency and swallowing difficulties can lead to consumption of a narrow, nutritionally imbalanced diet in the aged population [6–8]. Moreover, older adults have less money to spend on food.

All the above mentioned factors impact on the nutritional status of older adults, contributing to age-associated disorders including dysregulation of the immune system [9,10]. In fact, aging is associated with a declined immune function, a process known as immunosenescence, that negatively impacts on the capacity to properly respond to immune challenges thus contributing to the increased susceptibility of older persons to infections, poor vaccine efficacy and progressive development of low-grade, chronic inflammatory status [10–12].

Since a variety of bioactive dietary components have been shown to affect the immune system, an appropriate nutritional intervention may be a promising approach to counteract the impaired immune function occurring with aging [13,14]. Enriching staple or widely consumed foods can be a simple strategy to increase the intake of such components. Bread is an important food in the daily diet of several populations around the world. It is generally produced from refined white flour that lacks the nutrients, fibre and bioactive components present in the bran, but other ingredients can be added to increase the nutritional value of bread without altering its appearance and nature.

Lentils have been gaining increasing interest in the development of healthy and functional foods, due to the fact of their nutritional properties [15–20]. The existing varieties of lentils vary in colour, size and texture, but they all have a low level of antinutrients and a mild taste [21]. Lentils contain 28.7%–31.5% protein, which is considerable among legumes, and provide the essential amino acids lysine and leucine [22]. They are a valuable source of dietary fibre, mainly the insoluble component, but also the soluble one [19]. Dietary fibres provide many health benefits, such as lowering serum levels of LDL cholesterol, glucose and blood pressure, reducing constipation and other intestinal disorders and preventing intestinal cancer [23]. Moreover, the soluble fibres of lentils contain nutritionally significant amounts of prebiotic molecules, such as galacto-oligosaccharides (GOS) and fructo-oligosaccharides (FOS), that are known to selectively stimulate the growth and/or activity of some beneficial bacteria in the colon, having the potential to improve host health, such as several *Bifidobacteria* and lactobacilli strains [24–26]. Finally, lentils are reported to have a high content of phenolic compounds and to show a high antioxidant activity [27]. Actually, phytochemicals, and among them phenolic compounds, are known to have a major impact on health, since they can provide therapeutic benefits including prevention and/or treatment of diseases and physiological disorders [28]. Amongst the lentil varieties, red lentils distinguish themselves for being an important source of proteins, fibre and particularly of bioactive substances [29,30].

A recent study from our laboratory showed that red lentil flour can be blended with wheat flour up to 24% to produce bread with good volume, pleasant texture and taste [31]. We thus engaged in further studies, which are reported in the present paper, to describe the nutritional profile of our 24% red lentil bread and to get some insight into the in vivo effect of its consumption, with particular regard to the aging condition. The bread nutritional profile was described by analysing proteins, amino acids, lipids, soluble and insoluble dietary fibre, resistant starch, total polyphenols and specifically lignans, which is an interesting group of polyphenols present in pulses; in addition, its antioxidant power was measured by the ferric reducing antioxidant power (FRAP) assay.

The same bread was chosen for an in vivo experiment with aged mice, used as a vulnerable animal model, to evaluate if a substitution of common wheat bread with this special wheat–legume bread could counteract the immune decline typical of older adults. The immune response was mainly assessed at the intestinal level, since the mucosal immune system, which is known to be also impaired in the older adults [32], represents the first line of contact with ingested antigens and molecules reaching the intestinal lumen. Some parameters, namely, serum cytokines and intraepithelial lymphocyte immunophenotype were analysed, as they represent markers of systemic and intestinal inflammatory status, respectively.

2. Materials and Methods

2.1. Flours and Bread Preparation

Commercial wheat flour ("0" type according to the Italian flour classification, Horeca brand) and commercial dehulled red lentils (Select, San Giuseppe Vesuviano, Napoli, Italy) were purchased from the market.

The wheat flour had a moisture level of 12.8% (International Association for Cereal Science and Technology (ICC) standard 110/1 [33]), ash 0.63% d.m. (indicated on the product label), total protein of 10.5% d.m. (product label), lipids of 0.8% d.m. (product label) and total dietary fibre of 3.2% d.m. of which 2.1% was insoluble and 1.1% soluble (measured according to Lee et al. [34] using a reagent kit (K-TDFR, Megazyme Int., Wicklow, Ireland)).

Red lentils were ground in a refrigerated laboratory mill (M20, Janke and Kunkel Ika Labortechnik, Staufen, Germany) (a cutting/impact mill with no sieve, operating at a speed of 20,000 rpm for 2 min) to produce a very homogeneous flour that had a moisture level of 10.3% (ICC standard 110/1 [33]), ash content of 2.39% dry matter (d.m.) (ICC standard 104/1 [33]), total protein of 24.6% d.m. (product label), lipids of 1.3% d.m. (product label) and total dietary fibre content of 17.1% d.m. of which 15.2% was insoluble and 1.9% soluble (measured according to Lee et al. [34] using a reagent kit (K-TDFR, Megazyme Int., Wicklow, Ireland)).

A blend was prepared by mixing wheat flour with red lentil flour in the proportions of 76% and 24%, respectively. These proportions were chosen according to the results of Turfani et al. [31], who determined the maximum amount of red lentil flour that could be added to wheat flour in order to avoid technical problems during bread making, such as excessive dough sticking, poor dough rheological properties and bread with unacceptably low volume, poor texture and excessive legume flavour.

The bread formulation was kept simple in order to study the nutritional properties of bread produced from the flour blend without additives. Loaves of bread were produced from wheat flour (wheat bread) and from a wheat–lentil flour blend by adapting the ICC standard method No. 131 [33] because solution 1 was not used, thus reducing sugar and eliminating ascorbic acid from the ingredients (the same adapted method was used in References [20,31]). Thus, 1000 g of flour blend were weighted at 14% m.b. and mixed with 15 g salt in the mixer bowl; the optimum water amount (previously determined by the Brabender Farinograph according to ICC Standard 151/1 [33]) was added to the flour blend, except for the small amount required to activate yeast; compressed baker's yeast (18 g) was activated in 72 g of 5% sucrose solution (containing 68.4 g water and 3.6 g sucrose) at 35 °C for 10 min, then added to the flour blend. The dough was mixed for 10 min in a planetary bread mixer (Quick 20 by Sottoriva, Marano, Italy), then the dough temperature was checked (27 ± 1 °C) and the dough was fermented for 30 min in a fermentation cabinet at 30 °C with 85% relative humidity. After fermentation, the dough was scaled in four equal pieces, which were placed in baking tins and proofed for 50 min at 30 °C with 85% relative humidity, then baked for 30 ± 2 min at 220 °C in a convection/steam oven. The bread volume was determined within 20 ± 4 h by the rapeseed displacement method (AACCI Method 10-05.01) [35].

Bread for mouse feeding was baked all together at the beginning of the experiment to prevent variability due to the different preparation conditions, divided in aliquots sufficient for weekly diet preparation and frozen. Bread aliquots were thawed at room temperature at the moment of diet preparation.

2.2. Chemicals and Standards for Bread Analysis

The solvents used (i.e., acetone, diethyl ether, ethanol, ethyl acetate, methanol, n-hexane) were of HPLC or analytical grade and were purchased from Carlo Erba (Milan, Italy). Reagents were of the highest available purity. Hydrochloric acid 35%, formic acid 99%, glacial acetic acid, sulphuric acid 96%, tartaric acid, boric acid, sodium hydroxide, sodium hydroxide 32% solution, tris(hydroxymethyl)-aminomethane (TRIS), Folin–Ciocalteu reagent, sodium carbonate 20% solution, iron (II) sulphate heptahydrate (99%) and iron (III) chloride hexahydrate (97%–102%) were purchased from Carlo Erba. Kjieltabs (CuSO4/K2SO4), sulphuric acid solution 0.1 N and hydrogen peroxide 30% were purchased from VWR International PBI (Milan, Italy). Sodium citrate dihydrate, sodium acetate trihydrate, sodium chloride, glacial acetic acid, 2-metoxyethanol, ninhydrin were purchased from Merck-BDI (Darmstadt, Germany). Tin (II) chloride dehydrate, sodium acetate trihydrate 99%, MES (2(N-morpholino)-ethanesulpohonic acid), trolox

(6-hydroxy-2,5,7,8-tetramethylchroman-2-carboxylic acid), TPTZ (2,4,6-Tris(2-pyridyl)-s-triazine) and Helix Pomatia μ-glucuronidase/sulphatase S9626–10KU Type H-1, 0.7 G solid, 14,200 units/g solid were purchased from Sigma–Aldrich (Milan, Italy). Standards were of the highest available grade: Amino acids standards and gallic acid monohydrate were purchased from Sigma–Aldrich, whereas isolariciresinol, secoisolariciresinol, lariciresinol and pinoresinol were from Chemical Research (Rome, Italy). Ultra-pure water was produced by using in sequence a Millipore Elix 5 system and a Millipore Synergy 185 system (Millipore, Molsheim, France).

2.3. Proximate Composition, Amino Acids, Total Polyphenols, Lignans and Antioxidant Properties of Bread

Moisture, proteins (conversion factor 6.25 for legume flours and 5.70 for wheat flour), lipids and ash were determined by standard ICC methods 110/1, 105/2, 136, 104/1, respectively [33]. Soluble (SDFs), insoluble (IDFs) and total (TDFs) dietary fibres were determined according to Lee et al. [32] using a reagent kit (K-TDFR, Megazyme Int., Wicklow, Ireland). Available carbohydrates were calculated by difference. Resistant starch was determined according to AACC Method 32-40.01 [31] by means of a reagent kit (RSTAR, Megazyme); however, the results of all determinations were below the limit of detection (2%) and they are not shown in the tables.

Amino acids were determined according to Spackman, Stein and Moore [36] using a Beckman System Gold 126 amino acid analyser (Beckman Coulter Inc., Brea, CA, USA) equipped with a Beckman Spherogel IEX High-Performance Sodium column P/N 727450 and a Beckman UV detector with ninhydrin reactor. The samples were hydrolysed in hydrochloric acid 6 M under vacuum in sealed tubes at 105 °C for 24 h. For the determination of valine and isoleucine, the hydrolysis lasted 72 h. For cysteine and methionine, the samples were oxidised by oxygen peroxide and formic acid (88%) at 0 °C for 4 h at first, then the reagents were removed by evaporation under vacuum and the residue was hydrolysed in hydrochloric acid 6 M at 105 °C for 22 h. After hydrolysis and removal of the excess HCl, residues were re-dissolved in citrate buffer 0.2 M and injected.

Total polyphenols (TPCs) were extracted from samples as described by Durazzo et al. [30] in two separate fractions. Free polyphenols were extracted in methanol/water 1:1 and acetone/water 3:7. The residue was treated with hot sulphuric acid in methanol in order to free the hydrolysable polyphenols. The polyphenol content in the aqueous–organic extract and in the hydrolysed residue was determined by means of the Folin–Ciocalteau reagent [37], by measuring absorbance at 760 nm and using gallic acid as a standard.

For the analysis of lignans, samples were preliminarily defatted with hexane and diethyl ether for 8 h in a Soxhlet apparatus. The lignans were extracted and analysed by High Performance Liquid Chromatography (HPLC) as in Durazzo et al. [30]. The HPLC analyses were performed with a 50 μL extract using an ESA-HPLC system (ESA, Chelmsford, MA, USA) consisting of an ESA Model 540 autoinjector, an ESA Model 580 solvent delivery module with two pumps, an ESA 5600 eight channels coulometric electrode array detector and the ESA CoulArray operating software which controlled all the equipment and carried out data processing. A SUPELCOSIL LC-18 column (25 cm × 4.6 mm, 5 μm) with a Perisorb Supelguard LC-18 (Supelco, Milan, Italy) was used. Isolariciresinol, lariciresinol, secoisolariciresinol and pinoresinol were detected and quantified.

The antioxidant properties were determined by means of the FRAP assay according to Durazzo et al. [30].

2.4. In Vivo Experiments: Experimental Design, Animals and Diets

The Balb/c aged mice (20 months old) were kept at 23 °C with a 12 h light–dark cycle and fed ad libitum with standard laboratory diets. Mice had free access to food and water. Body weight and food intake were recorded every week and every other day, respectively. After one week of adaptation, animals were randomly divided into three groups (6 animals per group), receiving three different diets for two months (60 days): One group was fed a standard control diet (control group, 20% casein, Laboratorio Dottori Piccioni, Gessate, Milan, Italy), one group was fed the wheat bread containing diet (wheat bread group), and a third group was fed the wheat–lentil bread containing diet (wheat–lentil bread group). The standard control diet was prepared using as reference the AIN-93M formulation [38]. The bread containing diets were appropriately balanced and were isocaloric in respect to the control diet. At the end of the experimental periods, animals were fasted for 16 h, anesthetised with intraperitoneal injection of pentobarbital (10 mg/kg) and sacrificed. Blood was drawn via cardiac puncture, whereas small intestine and colon were excised and immediately placed in cold phosphate buffered saline (PBS). The animal experiments were carried out in strict accordance with the recommendation of the European Guidelines for the Care and Use of Animals for Research Purposes. All experimental procedures complied with the Animal Care and Use Committee of the CREA—Research Centre for Food and Nutrition—and were approved by the National Health Ministry, General Direction of Animal Health and Veterinary Drugs (agreement number 0006828/03/02/2014). All efforts were made to minimise the suffering of the animals.

2.5. Intraepithelial Lymphocytes (IELs) Preparation

The intraepithelial lymphocytes (IELs) were prepared from jejunum and colon. Briefly, intestines were placed on ice in 10 mL RPMI-1640 medium (Sigma–Aldrich, Milan, Italy), washed twice with cold PBS, longitudinally opened and cut into small size pieces. Intestinal pieces were washed in Hank's balanced salt solution (HBSS) and stirred twice for 45 min at 37 °C in an orbital shaker in HBSS added with 100 g/L foetal calf serum (FCS, Euroclone, Milan, Italy), 1 × 105 U/L penicillin, 100 mg/L streptomycin, 1 mM ethylendiamin-tetraacetic acid (EDTA), 5 mM Hepes, 1 mM dithiothreitol. The solution was passed through 100 and 40 μm nylon cell strainers (BD Falcon, Milan, Italy) and centrifuged at 650× *g*. The IELs were isolated from enterocytes by discontinuous 440/670 g/L Percoll gradient (PercollTM, GE Healthcare, Milan, Italy) in RPMI-1640 medium, and centrifuged at 650× *g* for 25 min.

2.6. Flow Cytometry Analysis of IELs Subpopulations

The following monoclonal antibodies were used for lymphocyte surface staining: Fluorescein isothiocyanate (FITC) anti-CD3 (clone 17.12), phycoerythrin (PE) anti-CD4 (clone GK1.5), phycoerythrin–cyanine 5 (PE-Cy5) anti-CD8 (clone 53-67), PE anti-CD19 (clone ID3), peridinin–chlorophyll-protein (PerCP) anti-CD45 (clone 30-F11), PE anti-TCR γδ (clone GL3), PE-Cy5 anti-TCR αβ (clone H57-597) and anti-CD16/CD32 (clone 2.4G2) (BD Pharmingen, Milan, Italy). Each antibody was previously titrated to determine the optimal concentration for maximal staining. The IELs (1 × 106 cells) were pre-incubated for 20 min with anti-CD16/CD32 to block Fc receptors. Cells were then washed and labelled with the appropriate mixture of antibodies for 30 min, centrifuged at 650× *g* and resuspended in FacsFlow (BD Biosciences, Milan, Italy). Flow cytometry analysis was performed using a FACSCalibur instrument (BD Biosciences). To exclude dead/dying cells and, therefore, non-specific antibody-binding cells, lymphocytes were gated according to forward and side scatter. The percentage of B and T lymphocytes was calculated on leukocyte (CD45+) gate, whereas the CD4+, CD8+ and CD4+CD8+ subsets, as well as αβ and γδ lymphocytes, were calculated on T lymphocyte (CD3+) gate. At least 10,000 events were acquired. Data were analysed using CellQuest software (BD Biosciences).

2.7. Analysis of Inflammatory Status in Mice Intestine

Small parts of jejunum and colon (1 cm) were immediately washed in cold PBS to remove stools and frozen in liquid nitrogen. To evaluate the inflammatory status of intestine, frozen tissues were weighted and homogenised in cold radioimmunoprotein (RIPA: 20 mM Tris-HCl pH 7.5, 150 mM NaCl, 0.1% Sodium Dodecyl Sulfate (SDS), 1% Na deoxycholate, 1% Triton X- 100) assay buffer supplemented with 1 mM phenylmethylsulphonyl fluoride, protease inhibitor cocktail (Complete Mini, Roche, Milan, Italy) and phosphatase inhibitor cocktail (PhosSTOP, Roche). Protein concentration was measured by the Lowry assay. Intestinal homogenates (50 μg total proteins) were dissolved in sample buffer (50 mM Tris-HCl, pH 6.8, 2% SDS, 10% glycerol, 100 g/L bromophenol blue, 10 mM beta-mercaptoethanol), heated for 5 min, fractionated by SDS-polyacrylamide gel (4–20% gradient) electrophoresis and transferred to 0.2 μm nitrocellulose filters (Trans-Blot Turbo, Biorad, Milan, Italy). Membranes were incubated with the following primary antibodies: Rabbit polyclonal anti-mouse NF-kB p65 and P-p65. Proteins were detected with horseradish peroxidase conjugated secondary antibodies (Cell Signaling Technology, Danvers, MA) and enhanced chemiluminescence reagent (ECL kit LiteAblot Extend, Euroclone, Milan, Italy), followed by analysis of chemiluminescence with the charge-coupled device camera detection system Las4000 Image Quant (GE Health Care Europe GmbH, Milan, Italy). The expression of the P-p65 proteins was normalised to their corresponding unphosphorylated forms.

2.8. Cytokine Secretion in Mice Serum

Blood samples were collected in test tubes, centrifuged (2000× g for 10 min at 4 °C) and the supernatant (serum) was stored at −80 °C until further analysis. The levels of cytokines and chemokines were analysed using Bio-plex/Luminex technology (mouse magnetic Luminex screening assay, Labospace, Milan, Italy). Briefly, Luminex multi-analyte profiling is a multiplexing technology allowing simultaneous analysis of up to 500 bioassays from a small sample volume. The following cytokines and chemokines were simultaneously detected in 50 μL undiluted samples: Tumour necrosis factor (TNF)-alpha, granulocyte macrophage-colony stimulating factor (GM-CSF), regulated upon activation normal T cell expressed and secreted (RANTES), interleukin (IL)-23, IL-17, IL-10, IL-12 and IL-6.

2.9. Presentation of Results and Statistics

Proximate composition, dietary fibre and lignan analyses were performed in triplicate whereas four replicates were used for polyphenols and FRAP. Mean values and percent coefficient of variation (%CV) are reported, together with the significance level of Student's t-test between wheat bread and wheat–lentil bread. Amino acids were analysed by a single determination without replicates. Calculations were performed by means of Microsoft Excel and PAST statistical package, version 2.17c [39].

For the results of in vivo experiments, values in graphs and tables represent means and %CV. Statistical significance was evaluated by one-way ANOVA followed by post-hoc Tukey's honestly significant difference (HSD) test. Normal distribution and homogeneity of variance of data were previously verified with appropriate statistical tests. Differences with p-values < 0.05 were considered significant. Statistical analysis was performed with the PAST statistical package.

3. Results

Table 1 shows the proximate composition of wheat bread and of the wheat–lentil bread. The two breads did not significantly differ in their moisture content (38.9% as is basis and 40.0% as is basis for the wheat bread and the wheat–lentil bread, respectively), whereas significant differences were observed for protein (6.4% as is basis and 8.3% as is basis for the wheat and the wheat–lentil bread, respectively), ash (0.39% as is basis and 0.63% as is basis for the wheat and the wheat-lentil bread, respectively) and IDF (1.6% as is basis and 3.1% as is basis for the wheat and the wheat-lentil bread, respectively).

Table 1. Proximate composition of wheat and wheat–lentil bread (76% wheat flour/24% red lentil) #.

Bread	Moisture%		Protein%		Fat%		Ash%		IDF §%		SDF §%		TDF §%		Available Carbohydrates (by Difference)%	
	Mean ns	CV	Mean **	CV	Mean ns	CV	Mean **	CV	Mean *	CV	Mean ns	CV	Mean ns	CV	Mean ns	CV
Wheat bread	38.9	0.8%	6.4	16%	1.0	0%	0.39	0%	1.6	13%	1.0	20%	2.6	15%	50.8	1.6%
Wheat–lentil bread	40.0	0.8%	8.3	12%	0.9	3%	0.63	0.02%	3.1	16%	1.5	60%	4.6	28%	45.5	4.0%

Values are the mean of three replicates, on wet basis; § IDF—insoluble dietary fibre; SDF—soluble dietary fibre; TDF—total dietary fibre; *, **, ns significance (t-test) of difference among the two samples: * significant at $p < 0.05$; significant ** at $p < 0.01$; ns not significant.

The content of 17 amino acids in mg/100 g proteins (eight essentials, tryptophan was not determined, plus alanine, arginine, aspartic acid, cysteine, glutamic acid, glycine, proline, serine and tyrosine) in both wheat and wheat–lentil bread is reported in Table 2. The main differences observed were aspartic acid (4.20 and 6.05 for the wheat and the wheat–lentil bread, respectively), glutamic acid (39.75 and 34.36 for the wheat and the wheat–lentil bread, respectively), proline (9.90 and 8.39 for the wheat and the wheat–lentil bread, respectively), lysine (2.18 and 3.30 for the wheat and the wheat–lentil bread, respectively) and arginine (3.89% and 4.91% for the wheat and the wheat–lentil bread, respectively).

Table 2. Amino acid composition of wheat and wheat–lentil bread (mg/100 g proteins) [#, §].

Sample	Aspartic Acid	Threonine	Serine	Glutamic Acid	Proline	Glycine	Alanine	Cystine	Valine
Wheat bread	4.20	2.82	4.98	39.75	9.90	3.67	3.03	1.89	4.53
Wheat–lentil bread	6.05	3.01	5.04	34.36	8.39	3.79	3.27	1.60	4.84
Sample	**Methionine**	**Isoleucine**	**Leucine**	**Tyrosine**	**Phenylalanine**	**Histidine**	**Lysine**	**Arginine**	**NH_4**
Wheat bread	1.36	4.07	7.08	2.77	4.88	2.28	2.18	3.89	5.12
Wheat–lentil bread	1.13	4.27	7.21	2.76	4.92	2.41	3.30	4.91	4.42

[#] Amino acids were analysed as a single determination without replicates. [§] Tryptophan was not analysed.

Data on total polyphenols content (TPC) (both in the aqueous organic extract and in the hydrolysable residue), four lignans content—namely, isolariciresinol, lariciresinol, secoisolariciresinol, pinoresinol—and the antioxidant power measured by the FRAP (both in the aqueous organic extract and hydrolysable residue) in our experimental wheat and wheat–lentil bread are reported in Table 3.

With regards to TPC, significant differences were observed between the wheat bread and the wheat–lentil bread both in the aqueous organic extract and the hydrolysable residue with values of 59.4 and 250.0 mg GAE/100 g d.m. in the aqueous organic extract of wheat and wheat–lentil bread, respectively, and higher values of 411.8 and 689.1 in the hydrolysable residue of the same samples.

With regards to the content of the four determined lignans, lariciresinol and pinoresinol were not detectable in the wheat bread whereas they reached 45.2. and 27.3 μg/100g d.m., respectively, in wheat–lentil bread. Significant differences between the two types of bread were observed for isolariciresinol (2.4 and 66.5 μg/100g d.m. for wheat and wheat–lentil bread, respectively) and for secoisolariciresinol (4.5 and 7.0 μg/100 g d.m. for wheat and wheat–lentil bread, respectively). Significant differences were also observed in the FRAP values of the aqueous organic extract and the hydrolysable residue of both types of bread which were higher for lentil bread in both cases (21.9 versus 6.4 and 49.7 versus 21.1 μmoL/g d.m., respectively).

Table 3. Total polyphenols, lignans and FRAP of wheat and wheat–lentil bread [#].

Bread	TPC (mg GAE/100 g d.m.)				Lignans (µg/100 g d.m.)								FRAP (µmol/g d.m.)			
	Aqueous–Organic Extract		Hydrolysable Residue		Isolariciresinol		Lariciresinol		Secoisolariciresinol		Pinoresinol		Aqueous–Organic Extract		Hydrolysable Residue	
	Mean **	CV	Mean **	CV	Mean *	CV	Mean	CV	Mean **	CV	Mean	CV	Mean **	CV	Mean **	CV
Wheat bread	59.4	1.5%	411.8	0.3%	2.4	8.3%	n.d.		4.5	4.4%	n.d.		6.4	3.1%	21.1	12.3%
Wheat–lentil bread	250.0	2.0%	689.1	4.9%	66.5	18.6%	45.2	12.2%	7.0	4.3%	27.3	18%	21.9	12.8%	49.7	7.8%

[#] Values are means of four to seven replicates; GAE—gallic acid equivalent; d.m.—dry matter; TPC—total phenols content; FRAP—ferric reducing antioxidant power; n.d.—not detectable. *, ** Significance (t-test) of differences among the two samples: Significant ** at $p < 0.01$ and * at $p < 0.05$.

Data on the composition of the diets which were given to the aged mice are reported in Table 4.

Table 4. Diets composition.

Component	Control (g/kg)	Wheat Bread (g/kg)	Wheat–Lentil Bread (g/kg)
Bread		465.7	465.7
Maize starch	465.7	66.4	92.6
Casein	140.0	88.8	74.8
Maltodextrins	155.0	155.0	155.0
Sucrose	100.0	100.0	100.0
Soya oil	40.0	35.6	35.4
Cellulose	50.0	39.2	27.2
Saline mix	35.0	35.0	35.0
Vitamin mix	10.0	10.0	10.0
L-cystine	1.8	1.8	1.8
Choline chloride	2.5	2.5	2.5
TBHQ #	0.008	0.008	0.008

tert-Butylhydroquinone.

Table 5 reports the data relative to mice initial (i.e., at the beginning of treatment) and final (i.e., at the end of treatment) body weight, as well as daily food intake. No significant differences were observed among the three groups in body weight nor in food intake.

Table 5. Body weight and daily food intake of control, wheat bread and wheat–lentil bread fed mice *.

Diet	Initial Body Weight (g)		Final Body Weight (g)		Food Intake (g/day)	
	Mean	CV	Mean	CV	Mean	CV
Control	24.0	8.3%	25.0	9.6%	3.6	22.2%
Wheat bread	25.5	8.2%	25.5	12.9%	3.4	23.5%
Wheat–lentil bread	24.0	15.0%	25.7	5.8%	3.7	24.3%

* Data represent means and %CV of 6 mice per group.

Among all the analysed cytokines and chemokines in serum, only three resulted at detectable levels: The anti-inflammatory IL-10, the pro-inflammatory IL-17 and the GM-CSF chemokine. Interleukin-10 significantly decreased in the wheat and wheat–lentil bread-treated animals as compared to control, whereas no significant differences were observed in IL-17 and GM-CSF levels among the three groups (Table 6).

Table 6. Cytokine serum secretion #.

Diet	Cytokine (pg/mL)					
	IL-17		IL-10		GM-CSF	
	Mean	CV	Mean	CV	Mean	CV
Control	25.79	4.96%	13.49	55.75%	3.02	10.26%
Wheat bread	23.01	10.30%	4.81 *	64.66%	2.50	22.40%
Wheat–lentil bread	23.52	1.0%	6.73 *	26.60%	2.65	3.77%

Data represent means and %CV of 6 mice per group; * $p < 0.05$ versus control.

The IELs subpopulation percentages in jejunum (panel A) and colon (panel B) of mice fed control, wheat bread or wheat–lentil bread diets are presented in Figure 1. Histograms show a significant increase of cytotoxic T cell (CD3+CD8+) percentages in the jejunum of mice fed both types of bread compared to the control diet, whereas the percentage of total T cells (CD3+CD45+) were reduced in mice fed wheat bread compared to control and wheat–lentil bread. In the colon, only a significant increase of B cell (CD19+CD45+) percentages was observed in mice fed wheat–lentil bread. No differences in the percentages of other lymphocyte subpopulations were observed among the groups.

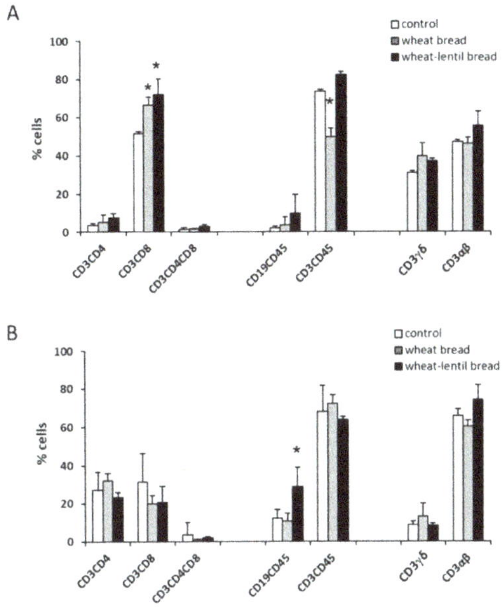

Figure 1. Intraepithelial lymphocyte (IEL) subpopulations in the jejunum (**A**) and colon (**B**) of mice fed a control, a wheat bread and a wheat–lentil diet measured by flow cytometry. (The percentage of B and T lymphocytes was calculated on leukocyte (CD45+) gate, whereas the CD4+, CD8+ and CD4+CD8+ subsets, as well as $\alpha\beta$ and $\gamma\delta$ lymphocytes, were calculated on T lymphocyte (CD3+) gate). Data represent the means ± SD of 6 mice. * $p < 0.05$ versus control.

Western blot analysis of the phosphorylated form of the p65 subunit of NF-kB in the jejunum and colon of mice did not show any significant difference among groups, indicating that the treatment with wheat and wheat–lentil bread did not induce an inflammatory status in the mice intestine (data not shown).

4. Discussion

As expected, the proximate composition of the wheat and the wheat–lentil bread mirrored the proximate composition of the flours of origin (see the Materials and Methods section and Table 1). In fact, the wheat–lentil bread contained 30% more proteins than wheat bread, it had an almost double ash content, therefore a higher level of minerals in general, together with an almost double amount of total dietary fibre, especially the insoluble component. Moreover, the lentil–wheat bread contained a lower amount of available carbohydrates than wheat bread.

Besides having a higher protein content, wheat–lentil bread had a more balanced amino acid profile than wheat bread (Table 2). Indeed, the amino acid profiles of wheat and lentils are complementary. For example, lysine is abundant in lentils, whereas sulphur amino acids are present in higher amounts in wheat. Lentil proteins are, in fact, mainly constituted by globulins and albumins [40] and, thus, have a different composition from wheat proteins, which are mainly constituted of prolamins and glutelins. The presence of the lentil flour increases the level of almost all the essential amino acids in bread (Table 2).

The lower amount of available carbohydrates in wheat–lentil bread is due to the fact that lentils contain less starch than wheat (about 40–45%, [18]); it is also reported that legume starch has a higher fraction of amylose than wheat (about 35%, [41]).

Regarding dietary fibre, both wheat and wheat–lentil breads contain only a small amount of the soluble component, around 1%; however, the soluble fibre of lentils is reported to contain beta glucans [18]. Beta-glucans are very interesting from a functional point of view, because they are known to induce a variety of physiological effects with a positive impact on health, acting in particular through immunomodulatory pathways, that can suppress cancer proliferation, lower cholesterol levels and thus reduce the risk for cardiovascular disease [42,43].

The wheat–lentil bread was richer in phenolic substances, in particular those present in the aqueous organic extract, than wheat bread and this is the reason why it also had better antioxidant properties (Table 3). The soluble free phenolics found in the extract come mainly from cellular vacuoles whereas the insoluble phenolics present in the residue are bound to other components mainly fibre. The hydrolysable bound phenols represent the main polyphenol fraction in both bread types (between 73% and 87% of TPC). The literature reports that significant amounts of phenolic compounds remain in the extraction residues, associated with the food matrix [44]. The phenolic molecules most frequently found in cereals are phenolic acids and flavonoids whereas in pulses we also find tannins [45]. Polyphenols in general, both the free and the bound ones, thanks to their antioxidant properties are considered to exert a protective effect on human health [46].

The lignans, secoisolariciresinol and isolariciresinol, were found in both breads (Table 3). However, the wheat–lentil bread not only contained higher amounts of these lignans, but also had additional lignan types and, in particular, lariciresinol and pinoresinol in the following order: Isolariciresinol > lariciresinol > pinoresinol > secoisolariciresinol. These results are in agreement with data on lignan content in legume flours reported by Durazzo et al. [30]. Literature data indicate flaxseed and sesame as major alimentary sources of lignans and rye and lentils as good sources [30,47]. Lignans are a large group of polyphenols of increasing interest because their intake has been related to beneficial health effects, including cancer and cardiovascular disease prevention [48]. In this regard it is interesting to report that in 2012 the research group of During et al. [49] published a paper to report on their investigation of whether plant lignans are taken up by intestinal cells and modulate the intestinal inflammatory response using the Caco-2 cell model. Their findings suggest that plant lignans can be absorbed and metabolised in the small intestine and, among the plants lignans tested, pinoresinol exhibited the strongest anti-inflammatory properties.

The antioxidant power as measured by FRAP was significantly higher in wheat–lentil bread than in wheat bread. The FRAP assay is a quick and sensitive way to measure the antioxidant capacity of biological samples. In both cases, the hydrolysable residue had a higher FRAP value than the aqueous–organic extract thus providing the major contribution to the total antioxidant power (from 69% to 77%); this matches the results of the total polyphenols content. Thus, a bread recipe where about one-quarter of the wheat flour is substituted by red lentil flour more than doubles the antioxidant capacity of bread (Table 3).

Concerning animal experiments, our first consideration was that no significant differences were observed in body weight and food intake among the groups of mice fed the control and the two types of bread diets; this indicates that the different diets had the same palatability for the mice, and that they did not impact on eating behaviour and appetite; and that all the differences observed in our sets of data were due to differences in the composition of the diets.

Analysis of the immune parameters in IELs isolated from jejunum showed an increase of cytotoxic T lymphocytes (CD3+CD8+) percentages in both wheat and wheat–lentil bread-treated animals, as compared to the control. Moreover, total T lymphocytes (CD3+CD45+) were significantly reduced in the wheat bread group and increased in the wheat–lentil bread group, compared to the control (Figure 1A). The IEL subpopulation's analysis in colon showed a significant increase of the B lymphocytes (CD19+CD45+) percentage in lentil bread-treated animals as compared to wheat bread and the control (Figure 1B). In this regard, we could hypothesise a role of the higher amount of β-glucans in the lentil bread while not ignoring that such compounds can increase the percentage of activated B lymphocytes and stimulate immune response [50,51].

We can say that the results of our study indicate a positive effect of wheat–lentil bread supplementation on the intestinal immune system of aged mice, as this supplementation was able to counteract some of the immune alterations typical of the older adults. In fact, aging is characterised by intrinsic changes in hematopoietic precursors that affect their proliferative potential, and this represents a key factor contributing to age-related decline in B- and T-cell production [52]. Thus, the increase of total T lymphocytes indicates a better immune response, and the increase of cytotoxic T lymphocytes suggests an improved capacity to respond to toxic agents and/or pathogens, that is known to be reduced in older adults. We can also hypothesise that the increase of B lymphocytes in the colon indicates a more efficient antibody response. In fact, it is well known that the antibody response is impaired in the older adults [53]. Moreover, it has been largely demonstrated that an antioxidants-containing diet may ameliorate lymphocyte response and protect immune cells from oxidative stress-induced apoptosis [54]. Besides polyphenols in general, the positive effects on the immune system in our specific case could also be ascribed to the significantly higher amount of the lignan isolariciresinol (27 times higher) and the presence of the lignan pinoresinol in wheat–lentil bread compared to wheat bread; these two lignans in particular have been shown to exert immunomodulatory and anti-inflammatory effects [49,55].

No significant differences were observed in the other analysed IELs subpopulations (Figure 1A,B).

Among all the analysed cytokines in serum, only IL-10 was significantly decreased in the wheat and wheat–lentil bread treated animals as compared to the control (Table 6). The role of IL-10 in older adults is controversial; while some studies report that IL-10 increased the inflammatory status, others indicate that this cytokine plays a key role as an anti-inflammatory factor [56,57]. It has also been reported that aging is associated with an increase of IL-10 that, together with other cytokines, could be considered as a marker of frailty [58,59].

5. Conclusions

It is increasingly coming to general attention that the aging population needs to eat appropriately to prevent and reduce all the health risks associated with this phase of human life. In other words, there is a need for tailored foods for aging people. Enriching staple or widely consumed foods can be a simple strategy to guaranty the intake of key nutrients able to have a beneficial effect on the negative aspects associated with aging such as the decline of the immune function. Based on previous studies done in our laboratory, we identified bread as a target food and red lentil flour as a raw material useful to add functionality to bread. We also identified technological constraints that allowed a maximum addition of 24% lentil flour.

For the purpose of this study, we baked two kinds of bread: A common wheat bread and a wheat–24% lentil flour bread. The chemical analysis of the bread components showed that the wheat–lentil bread had 30% more proteins than wheat bread coupled with a more balanced amino acid composition; it had an almost double mineral as well as total dietary fibre content, especially the insoluble component, double the amount of polyphenols, an interesting lignans content and more than double the antioxidant capacity. Thus, this wheat–lentil bread proved to be nutritionally richer and more functional than common wheat bread.

The in vivo effect of the consumption of wheat–lentil bread versus wheat bread on the immune response was studied by means of a murine model of aged mice. Analysis of the immune parameters in intraepithelial lymphocytes isolated from the mice intestine showed significant differences between the two types of bread indicating a positive effect of the lentil–wheat bread on the intestinal immune system. Cytokines in serum were also analysed. Considering that IL-10 is indicated as a frailty marker, we suppose that wheat and wheat–lentil breads in diets could have a positive effect on inflammatory status and improve the health status of aged mice.

This study clearly demonstrates that this is possible by substituting wheat flour with another suitable flour to manufacture a simple and well-accepted food, such as bread, which shows more functionality and is more tailored for the aging population than traditional, common bread with soft wheat only.

Author Contributions: M.C. planned and supervised the research, acquired funds and wrote the final manuscript, V.N. and V.T. performed all the bread analyses but the lignans which were performed by A.D. A.F. and M.R. designed the animal experimental setup and performed the experiments and analysed and interpreted the data. R.R. prepared the experimental diets and took care of the animals.

Funding: This research was carried out within the CERSUOM project financed by the Italian Ministry for Agricultural, Alimentary and Forest Policies (MiPAAF).

Acknowledgments: Grateful acknowledgements are due to Luigi Bartoli and Francesco Mellara for technical help in the bread preparation.

Conflicts of Interest: The authors declare no conflict of interest.

References

1. EUROSTAT. Your Key to European Statistics. 2019. Available online: https://ec.europa.eu/eurostat/data/database (accessed on 25 July 2019).
2. Morley, J.E. Undernutrition in older adults. *Fam. Pract.* **2012**, *29*, i89–i93. [CrossRef] [PubMed]
3. Starr, K.N.P.; McDonald, S.R.; Bales, C.W. Nutritional vulnerability in older adults: A continuum of concerns. *Curr. Nutr. Rep.* **2015**, *4*, 176–184. [CrossRef]
4. Lichtenstein, A.H.; Rasmussen, H.; Yu, W.W.; Epstein, S.R.; Russell, R.M. Modified MyPyramid for older adults. *J. Nutr.* **2008**, *138*, 5–11. [CrossRef] [PubMed]
5. Ziylan, C.; Haveman-Nies, A.; Kremer, S.; de Groot, L.C. Protein-enriched bread and readymade meals increase community-dwelling older adults' protein intake in a double-blind randomized controlled trial. *J. Am. Med. Dir. Assoc.* **2016**, *18*, 145–151. [CrossRef] [PubMed]
6. Boyce, J.M.; Shone, G.R. Effects of ageing on smell and taste. *Postgrad. Med. J.* **2006**, *82*, 239–241. [CrossRef] [PubMed]
7. Ney, D.M.; Weiss, J.M.; Kind, A.J.; Robbins, J. Senescent swallowing: Impact, strategies, and interventions. *Nutr. Clin. Pract.* **2009**, *24*, 395–413. [CrossRef] [PubMed]
8. Rémond, D.; Shahar, D.R.; Gille, D.; Pinto, P.; Kachal, J.; Peyron, M.A.; Dos Santos, C.N.; Walther, B.; Bordoni, A.; Dupont, D.; et al. Understanding the gastrointestinal tract of the elderly to develop dietary solutions that prevent malnutrition. *Oncotarget* **2015**, *6*, 13858–13898. [CrossRef]
9. Chandra, R.K. Nutrition and the immune system from birth to old age. *Eur. J. Clin. Nutr.* **2002**, *56*, S73–S76. [CrossRef]
10. Fülöp, T.; Dupuis, G.; Witkowski, J.M.; Larbi, A. The role of immunosenescence in the development of age-related diseases. *Rev. Investig. Clín.* **2016**, *68*, 84–91.
11. Castelo-Branco, C.; Soveral, I. The immune system and aging: A review. *Gynecol. Endocrinol.* **2014**, *30*, 16–22. [CrossRef]
12. Pera, A.; Campos, C.; López, N.; Hassouneh, F.; Alonso, C.; Tarazona, R.; Solana, R. Immunosenescence: Implications for response to infection and vaccination in older people. *Maturitas* **2015**, *82*, 50–55. [CrossRef] [PubMed]
13. Pae, M.; Meydani, S.N.; Wu, D. The Role of nutrition in enhancing immunity in aging. *Aging Dis.* **2012**, *3*, 91–129. [PubMed]
14. Vulevic, J.; Juric, A.; Walton, G.E.; Claus, S.P.; Tzortzis, G.; Toward, R.E.; Gibson, G.R. Influence of galacto-oligosaccharide mixture (B-GOS) on gut microbiota, immune parameters and metabonomics in elderly persons. *Br. J. Nutr.* **2015**, *114*, 586–595. [CrossRef] [PubMed]
15. Dalgetty, D.D.; Baik, B.K. Fortification of bread with hulls and cotyledon fibers isolated from peas, lentils and chickpeas. *Cereal Chem.* **2006**, *83*, 269–274. [CrossRef]
16. Borsuk, Y.; Arntfield, S.; Lukow, O.M.; Swallow, K.; Malcolmson, L. Incorporation of pulse flours of different particle size in relation to pita bread quality. *J. Sci. Food Agric.* **2012**, *92*, 2055–2061. [CrossRef] [PubMed]
17. Baik, B.K.; Han, I.H. Cooking, roasting and fermentation of chickpeas, lentils, peas and soybeans for fortification of leavened bread. *Cereal Chem.* **2012**, *89*, 269–275. [CrossRef]
18. Faris, M.A.I.E.; Takruri, H.R.; Issa, A.Y. Role of lentils (Lens culinaris L.) in human health and nutrition: A review. *Med. J. Nutr. Metab.* **2012**, *6*, 3–16. [CrossRef]

19. Aslani, Z.; Alipour, B.; Mirmiran, P.; Bahadoran, Z. Lentil's (Lens culinaris L.) functional properties in prevention and treatment of non-communicable chronic diseases: A review. *Int. J. Nutr. Food Sci.* **2015**, *4*, 15–20. [CrossRef]
20. Turfani, V.; Narducci, V.; Durazzo, A.; Galli, V.; Carcea, M. Technological, nutritional and functional properties of wheat bread enriched with lentil or carob flours. *LWT-Food Sci. Technol.* **2017**, *78*, 361–366. [CrossRef]
21. Nienke, M.; Lindeboom, T.; Baga, M.; Vandenberg, A.; Chibbar, R.N. Composition and correlation between major seed constituents in selected lentil (Lens culinaris. Medik) genotypes. *Can. J. Plant Sci.* **2011**, *91*, 825–835.
22. Roy, F.; Boye, J.I.; Simpson, B.K. Bioactive proteins and peptides in pulse crops: Pea, chickpea and lentil. *Food Res. Int.* **2010**, *43*, 432–442. [CrossRef]
23. Hefnawy, T.H. Effect of processing methods on nutritional composition and anti-nutritional factors in lentils (Lens culinaris). *Ann. Agric. Sci.* **2011**, *56*, 57–61. [CrossRef]
24. Johnson, C.R.; Thavarajah, D.; Combs, G.F., Jr.; Thavarajah, P. Lentil (*Lens culinaris* L.): A prebiotic-rich whole food legume. *Food Res. Int.* **2013**, *51*, 107–113. [CrossRef]
25. Gibson, G.R.; Hutkins, R.; Sanders, M.E.; Prescott, S.L.; Reimer, R.A.; Salminen, S.J.; Scott, K.; Stanton, C.; Swanson, K.S.; Cani, P.D.; et al. Expert consensus document: The International Scientific Association for Probiotics and Prebiotic s (ISAPP) consensus statement on the definition and scope of prebiotics. *Nat. Rev. Gastroenterol. Hepatol.* **2017**, *14*, 491–502. [CrossRef]
26. Micioni Di Bonaventura, M.V.; Cecchini, C.; Vila-Donat, P.; Caprioli, G.; Cifani, C.; Coman, M.M.; Cresci, A.; Fiorini, D.; Ricciutelli, M.; Silvi, S.; et al. Evaluation of the hypocholesterolemic effect and prebiotic activity of a lentil (*Lens culinaris* Medik) extract. *Mol. Nutr. Food Res.* **2017**, *61*, 1700403. [CrossRef] [PubMed]
27. Johari, A.; Arora, S.; Potaliya, M.; Kawatra, A. Role of Lentil (*Lens culinaris* L.) in human health and nutrition. *Ann. Agri-Bio Res.* **2015**, *20*, 291–294.
28. Bouchenak, M.; Lamri-Senhadji, M. Nutritional quality of legumes, and their role in cardiometabolic risk prevention: A review. *J. Med. Food* **2013**, *16*, 185–198. [CrossRef]
29. Wang, N.; Hatcher, D.W.; Toews, R.; Gawalko, E.J. Influence of cooking and dehulling on nutritional composition of several varieties of lentils (Lens culinaris). *Food Sci. Technol.* **2009**, *42*, 842–848. [CrossRef]
30. Durazzo, A.; Turfani, V.; Azzini, E.; Maiani, G.; Carcea, M. Phenols, lignans and antioxidant properties of legume and sweet chestnut flours. *Food Chem.* **2013**, *140*, 666–671. [CrossRef]
31. Turfani, V.; Narducci, V.; Mellara, F.; Bartoli, L.; Carcea, M. Technological properties and bread characteristics of soft wheat and red lentil flour blends. *Tec. Molit. Int.* **2017**, *68*, 52–72.
32. Mabbott, N.A.; Kobayashi, A.; Sehgal, A.; Bradford, B.M.; Pattison, M.; Donaldson, D.S. Aging and the mucosal immune system in the intestine. *Biogerontology* **2015**, *16*, 133–145. [CrossRef] [PubMed]
33. International Association for Cereal Science and Technology. *ICC Standard Methods*; Methods No. 104/1, 105/2, 110/1, 115/1, 131, 136; ICC: Vienna, Austria, 2003.
34. Lee, S.C.; Prosky, L.; DeVries, J.W. Determination of total, soluble and insoluble, dietary fibre in foods, enzymatic-gravimetric method, MES-TRIS buffer: Collaborative study. *J. Assoc. Off. Anal. Chem.* **1992**, *75*, 395–416.
35. AACC. *International Approved Methods of Analysis*, 11th ed.; Method 10-05.01 "Guidelines for Measurement of Volume by Rapeseed Displacement" and Method 32-40.01 "Resistant Starch in Starch Samples and Plant Materials"; AACC International: St. Paul, MN, USA, 2009.
36. Spackman, D.K.; Stein, W.H.; Moore, S. Chromatography of amino acids on sulfonated polystyrene resin: An improved system. *Anal. Chem.* **1958**, *30*, 1190–1196. [CrossRef]
37. Singleton, V.L.; Orthofer, R.; Lamuela-Raventós, R.M. Analysis of total phenols and other oxidation substrates and antioxidants by means of Folin-Ciocalteu reagent. *Methods Enzymol.* **1999**, *299*, 152–178.
38. Reeves, P.G. Components of the AIN-93 diets as improvements in the AIN-76A diet. *J. Nutr.* **1997**, *127*, 838S–841S. [CrossRef] [PubMed]
39. Hammer, Ø.; Harper, D.A.T.; Ryan, P.D. PAST: Paleontological Statistics Software Package for Education and Data Analysis. *Palaeontol. Electron.* **2001**, *4*, 1–9.
40. Boye, J.; Zare, F.; Pletch, A. Pulse proteins: Processing, characterization, functional properties and applications in food and feed. *Food Res. Int.* **2010**, *43*, 414–431. [CrossRef]
41. Joshi, M.; Aldred, P.; Panozzo, J.F.; Kasapis, S.; Adhikari, B. Rheological and microstructural characteristics of lentil starch-lentil protein composite pastes and gels. *Food Hydrocoll.* **2014**, *35*, 226–237. [CrossRef]

42. Earnshaw, S.R.; McDade, C.L.; Chu, Y.; Fleige, L.E.; Sievenpiper, J.L. Cost-effectiveness of maintaining daily intake of oat β-glucan for coronary heart disease primary prevention. *Clin. Ther.* **2017**, *39*, 804–818. [CrossRef]
43. Roudi, R.; Mohammadi, S.R.; Roudbary, M.; Mohsenzadegan, M. Lung cancer and β-glucans: Review of potential therapeutic applications. *Investig. New Drugs* **2017**, *35*, 509–517. [CrossRef]
44. Pérez-Jiménez, J.; Torres, J.L. Analysis of nonextractable phenolic compounds in foods: The current state of the art. *J. Agric. Food Chem.* **2011**, *59*, 12713–12724. [CrossRef] [PubMed]
45. Carcea, M.; Narducci, V.; Turfani, V.; Giannini, V. Polyphenols in raw and cooked cereals/pseudocereals/legume pasta and couscous. *Foods* **2017**, *6*, 80. [CrossRef] [PubMed]
46. Vitaglione, P.; Napolitano, A.; Fogliano, V. Cereal dietary fibre: A natural functional ingredient to deliver phenolic compounds into the gut. *Trends Food Sci. Technol.* **2008**, *19*, 451–463. [CrossRef]
47. Milder, I.E.; Arts, I.C.; van de Putte, B.; Venema, D.P.; Hollman, P.C.H. Lignan contents of Dutch plants foods: A database inscluding lariciresinol, pinoresinol, secoisolariciresinol and matairesinol. *Br. J. Nutr.* **2005**, *93*, 393–402. [CrossRef]
48. Adlercreutz, H. Lignans and human health. *Crit. Rev. Clin. Lab. Sci.* **2007**, *44*, 483–525. [CrossRef]
49. During, A.; Debouche, C.; Raas, T.; Larondelle, Y. Among plant lignans, pinoresinol has the strongest anti-inflammatory properties in human intestinal Caco-2 cells. *J. Nutr.* **2012**, *142*, 1798–1805. [CrossRef]
50. Dong, S.F.; Chen, J.M.; Zhang, W.; Sun, S.H.; Wang, J.; Gu, J.X.; Boraschi, D.; Qu, D. Specific immune response to HBsAg is enhanced by beta-glucan oligosaccharide containing an alpha-(1–>3)-linked bond and biased towards M2/Th2. *Int. Immunopharmacol.* **2007**, *7*, 725–733. [CrossRef]
51. Bobadilla, F.; Rodriguez-Tirado, C.; Imarai, M.; Galotto, M.J.; Andersson, R. Soluble β-1,3/1,6-glucan in seaweed from the southern hemisphere and its immunomodulatory effect. *Carbohydr. Polym.* **2013**, *92*, 241–248. [CrossRef]
52. Min, H.; Montecino-Rodriguez, E.; Dorshkind, K. Effects of aging on early B-and T-cell development. *Immunol. Rev.* **2005**, *205*, 7–17. [CrossRef]
53. Riley, R.L.; Blomberg, B.B.; Frasca, D. B cells, E2A, and aging. *Immunol. Rev.* **2005**, *205*, 30–47. [CrossRef]
54. Gollapudi, S.; Gupta, S. Reversal of oxidative stress-induced apoptosis in T and B lymphocytes by Coenzyme Q10 (CoQ10). *Am. J. Clin. Exp. Immunol.* **2016**, *5*, 41–47. [PubMed]
55. Xiao, P.; Huang, H.; Chen, J.; Li, X. In vitro antioxidant and anti-inflammatory activities of Radix Isatidis extract and bioaccessibility of six bioactive compounds after simulated gastro-intestinal digestion. *J. Ethnopharmacol.* **2014**, *157*, 55–61. [CrossRef] [PubMed]
56. Gao, S.; Shu, S.; Wang, L.; Zhou, J.; Yuan, Z. Pro-inflammatory and anti-inflammatory cytokine responses of peripheral blood mononuclear cells in apparently healthy subjects. *Nan Fang Yi Ke Da Xue Xue Bao* **2014**, *34*, 1589–1593. [PubMed]
57. Măluţan, A.M.; Drugan, T.; Ciortea, R.; Mocan-Hognogi, R.F.; Bucuri, C.; Rada, M.; Mihu, D. Serum anti-inflammatory cytokines for the evaluation of inflammatory status in endometriosis. *J. Res. Med. Sci.* **2015**, *20*, 668–674. [PubMed]
58. Teimourian, M.; Jafaraian, Z.; Hosseini, S.R.; Rahmani, M.; Bagherzadeh, M.; Aghamajidi, A.; Bijani, A.; Nooreddini, H.; Mostafazadeh, A. Both immune hormone IL-10 and parathormone tend to increase in serum of old osteoporotic people independently of 1, 25 dihydroxy vitamin D3. *Casp. J. Intern. Med.* **2016**, *7*, 283–289.
59. Langmann, G.A.; Perera, S.; Ferchak, M.A.; Nace, D.A.; Resnick, N.M.; Greenspan, S.L. Inflammatory Markers and Frailty in Long-Term Care Residents. *J. Am. Geriatr. Soc.* **2017**, *65*, 1777–1783. [CrossRef]

© 2019 by the authors. Licensee MDPI, Basel, Switzerland. This article is an open access article distributed under the terms and conditions of the Creative Commons Attribution (CC BY) license (http://creativecommons.org/licenses/by/4.0/).

foods

MDPI

Article

Gluten-Free Bread with Cricket Powder—Mechanical Properties and Molecular Water Dynamics in Dough and Ready Product

Przemysław Łukasz Kowalczewski [1,*], Katarzyna Walkowiak [2], Łukasz Masewicz [2], Olga Bartczak [3], Jacek Lewandowicz [4], Piotr Kubiak [5] and Hanna Maria Baranowska [2]

1 Institute of Food Technology of Plant Origin, Poznań University of Life Sciences, 60-624 Poznań, Poland
2 Department of Physics and Biophysics, Poznań University of Life Sciences, 60-637 Poznań, Poland
3 Students' Scientific Club of Food Technologists, Poznań University of Life Sciences, 60-624 Poznań, Poland
4 Chair of Production Engineering and Logistics, Poznan University of Technology, 60-965 Poznań, Poland
5 Department of Biotechnology and Food Microbiology, Poznań University of Life Sciences, 60-627 Poznań, Poland
* Correspondence: przemyslaw.kowalczewski@up.poznan.pl

Received: 2 June 2019; Accepted: 1 July 2019; Published: 3 July 2019

Abstract: Published data indicate that cricket powder (CP) is a good source of not only protein, fat and fiber, but also minerals. Due to the fact that this product naturally does not contain gluten, it is an interesting addition to the enrichment of gluten-free foods. This paper is a report on the results of starch substitution with CP (at 2%, 6% and 10%) on the properties of dough and bread. The rheology of dough and the texture of the final product were studied. While the changes caused in the dough by the introduction of CP were not pronounced, the bread obtained from it was characterized by significantly increased hardness and improved consistency. Analyses of water behavior at the molecular level with the use of ^{1}H Nuclear Magnetic Resonance (NMR) indicated that CP altered both the bound and bulk water fractions. Moreover, examination of water activity revealed a decreased rate of water transport in samples of bread that contained CP. These results indicate improved availability of water to the biopolymers of bread, which likely plays a role in shaping the textural properties of the product.

Keywords: gluten-free bread; edible insects; protein enrichment; rheology; texture; ^{1}H NMR; water behavior; water activity

1. Introduction

An increasing number of patients with celiac disease has led to increased interest in gluten-free (GF) products. Celiac disease is characterized by permanent gluten intolerance which, in turn, results in histopathological changes within the mucosa of the small intestine [1]. The only effective way to combat it is strict adherence to the GF diet [2,3]. It is estimated that about 1% of the population suffers from this disease [4–7]. Although the effectiveness of a gluten-free diet has not yet been proven for other diseases except celiac disease, it is also often recommended by doctors in other disease entities, such as non-celiac gluten sensitivity, Hashimoto's disease and irritable bowel syndrome [8]. Thus, the GF product market continues to grow.

Gluten is responsible for the retention of gases in dough, as well as for giving dough the right consistency. GF bread is characterized by structure and texture that is generally perceived as unattractive. In order to improve the properties of bread, including its aroma, additives, for example hydrocolloids, are used [9–12]. Moreover, GF bakery products are characterized by improper nutrient composition that results from the substitution of gluten containing flour with alternative starchy raw materials. Compared to traditional cereal products, GF breads have significantly lower nutritional value, especially in terms of decreased content of fiber, minerals and protein [9,13,14]. The additives used for the production of GF bread can supply the missing nutrients. Among such additives, edible insects can be distinguished.

In Africa, Latin America and Asia, edible insects have been known as a foodstuff for years [15,16]. As reported by the United Nations Food and Agriculture Organization, more than 1900 species of insects are eaten worldwide, including crickets, meal larvae, ants, grasshoppers and flies [17]. Research results published so far indicate that crickets, as well as cricket powder obtained from them, are a valuable source of protein, fat and minerals [18–20]. They also contain bioactive compounds [21,22]. Efforts have thus been undertaken to introduce them to the production of many food products [23–25]. To date, however, the impact of cricket powder (CP) on the characteristics of GF dough and the texture of GF bread has not been described. Therefore, the aim of the work was to assess the influence of cricket powder on the rheological properties of dough and the resulting texture of GF bread. Furthermore, water behavior in the tested bread samples was investigated with the use of low-field Nuclear Magnetic Resonance (NMR).

2. Materials and Methods

2.1. Materials

Corn starch was purchased from Glutenex (Sady, Poland), potato starch from PPZ Trzemeszno sp. z o.o. (Trzemeszno, Poland), guar gum from Limpio Chem LLP (Gujarat, India), pectin from Silvateam S.p.a. (via Torre, Italy), yeast from Lesaffre Polska (Wolczyn, Poland), sugar from Pfeifer & Langen Polska S.A. (Środa Wielkopolska, Poland), salt from Ciech Soda Polska S.A. (Janikowo, Poland) and rapeseed oil from ZT 'Kruszwica' S.A. (Kruszwica, Poland). The edible cricket powder was obtained from Crunchy Critters (Derby, United Kingdom). All chemicals and reagents used were of analytical grade.

2.2. Production of Bread

The recipe for reference gluten-free bread was as follows: 200 g corn starch, 50 g potato starch, 4.25 g guar gum, 4.25 g pectin, 15 g yeast, 5 g sugar, 4.25 g salt, 7.5 g rapeseed oil and 275 g distillated water [26]. Dough was prepared using the straight dough method. All the compounds, except oil, were mixed together with the use of KitchenAid mixer (model 5KPM5EWH, KitchenAid, Benton Harbor, MI, USA) for 2 min at a speed of 70 rpm, then oil was added, and mixing was continued for 6 min. Next, the dough was fermented in a fermentation chamber for 20 min (temperature 35 °C, relative humidity 85%) and punched. Each sample of dough was divided into two parts (280 g each) and placed in baking forms. The final fermentation was carried out for 15 min at 35 °C. Prepared dough was baked at 230 °C for 30 min (MIWE Michael Wenz GmbH, Amstein, Germany). Afterwards, the obtained breads were left at room temperature for 2 h to cool down, weighed and packed in polypropylene pouches. In the test samples, total starch was replaced with CP in three different quantities of 2%, 6% and 10%; the amounts of other components were unchanged. Reference dough and bread were denoted in the text as DB and RB, respectively. The dough samples containing cricket powder were named DCP2, DCP6, and DCP10 and the bread samples obtained from them were named BCP2, BCP6, and BCP10, respectively.

2.3. Rheological Properties of Dough

Viscoelastic properties were determined with the RheoStress1 rheometer (Haake Technik GmbH, Vreden, Germany) in controlled deformation mode (CD) with deformation set to 0.5%. Mechanical spectra were obtained within an angular velocity range of 0.1–100 rad·s^{-1}. The diameter parallel plate measurement geometrics (PP35 Ti) were 35 mm with a 1.0 mm gap. Complex viscosity (η^*), storage modulus (G′), and loss modulus (G″) were determined. The Ostwald de Waele equation (η*) and the power law equations (G′ and G″) were used to model the obtained spectra.

$$\eta^* = K^* \times \omega^{n^*-1}, \quad (1)$$

where η* is complex viscosity (Pa·s), K* is consistency index (Pa·s^n), ω is angular velocity (rad·s^{-1}) and n^* is flow behavior index (-).

$$G' = K' \times \omega^{n'}, \quad (2)$$

where G′ is storage modulus (Pa), K′ is the equation constant (Pa·s^n), ω is angular velocity (rad·s^{-1}), n' is the equation constant (-).

$$G'' = K'' \times \omega^{n''}, \quad (3)$$

where G″ is loss modulus (Pa), K″ is equation constant (Pa·s^n), ω is angular velocity (rad·s^{-1}), n'' is equation constant (-).

2.4. Texture Analysis

Texture profile analysis of bread was performed with a TA.XTplus texture analyzer (Stable Micro System Co. Ltd., Surrey, England) equipped with a 5 kg load cell [27]. Each sample was compressed twice with a cylindrical plunger probe with a 35 mm diameter. The test parameters were as follows: 10.0 mm s^{-1} pre-test speed, 5.0 mm s^{-1} test speed, 5.0 mm s^{-1} post-test speed, and 40% strain. Bread loaves were cut into slices (25 mm thick each and ends were discarded) and used to evaluate hardness, springiness, cohesiveness, chewiness and resilience. Texture analysis was repeated 15 times for each sample.

2.5. NMR Relaxometry

NMR measurements were performed according to Baranowska et al. [28]. Crumb or dough samples of 1.5 cm^3 were placed in measuring test tubes and sealed using Parafilm® (Bemis Company, Inc., Joplin, MO, USA). Measurements of the spin–lattice (T_1) and spin–spin (T_2) relaxation times were performed using a pulse NMR spectrometer MSL30 operating at 30 MHz (WL Electronics, Poznań, Poland). The samples were measured at 21.0 ± 0.5 °C. The inversion-recovery (180–t–90) [29] pulse sequence was used for measurements of the T_1 relaxation times. Distances between RF pulses (t) were changed within the range from 80 to 130 ms and the repetition time was 10 s. Each time, 32 free induction decay (FID) signals and 119 points for each FID signal were collected. Calculations of the spin–lattice relaxation time values were performed in CracSpin program using the 'spin grouping' approach. Marquardt's method of minimization was used for fitting multiexponential decays. Standard deviation was used to determine the accuracy of the analysis of relaxation parameters. Time changes of the current value of the FID signal amplitude in the employed frequency of impulses were described by the following formula:

$$M_z(t) = M_0\left\{1 - 2\exp\left(\frac{-t}{T_1}\right)\right\}, \quad (4)$$

where $M_z(t)$ is the actual magnetization value and M_0 is the equilibrium magnetization value.

Magnetization recovery was determined monoexponentially, which means that the system relaxes with one T_1 spin–lattice relaxation time. Measurements of the spin–spin (T_2) relaxation times were taken using the pulse train of the Carr–Purcell–Meiboom–Gill spin echoes ($\pi/2$–TE/2–$(\pi)_n$) [29]. The distance (τ) between 180 RF pulses amounted to 1 ms. The repetition time was 10 s. The number of spin echoes (n) amounted to 50. Eight accumulation signals were employed. To calculate the spin–spin relaxation time values, the authors applied adjustment of values of the echo amplitudes to the formula [30]:

$$M_{x,y}(\tau) = M_0 \sum_{i=1}^{n} p_i \exp\left[\frac{-\tau}{T_{2i}}\right], \quad (5)$$

where $M_{x,y}(\tau)$ is the echo amplitude, M_0 is the equilibrium amplitude, and p_i is the fraction of protons relaxing with the T_{2i} spin–spin time.

The calculations were performed using the dedicated software by application of the non-linear least-square algorithm. The accuracy of the relaxation parameters was estimated with standard deviation. The presence of two proton fractions was determined for all analyzed systems.

2.6. Measurements of Water Activity

Analyses of water activity a_w in the bread crumbs were conducted using a water diffusion and activity analyzer, ADA-7 (COBRABID, Poznań, Poland), with automatic recording of water evacuation from individual samples [31]. The thickness of the sample placed in the measurement chamber was 5 mm. Before the analysis, the temperature was stabilized at 21.0 ± 0.1 °C. The sample was then dried to the activity of 0.1000 ± 0.0005. The duration of each measurement was 1200 s. Water activity measurement results were used to describe water transport in breads with the use of the following phenomenological model [32]:

$$a_w(t) = a_r + \left(a_0 - a_p\right)e^{-V_D t} + \left(a_p - a_r\right)e^{-V_p t}, \quad (6)$$

where $a_w(t)$ is the temporary water activity value, a_0 is the initial water activity, a_p is the limit water activity (intermediate), a_r is the water activity at equilibrium condition (final), V_D is the transport rate, and V_p is the rate of the surface conduction.

2.7. Statistical Analysis

For every test, three independent measurements were taken, unless stated otherwise. One-way analysis of variance was performed independently for each dependent variable. Post-hoc Tukey HSD multiple comparison tests were used to identify statistically homogeneous subsets at $\alpha = 0.05$. Statistical analysis of the data was performed with Statistica 13 (Dell Software Inc., Round Rock, TX, USA) software.

3. Results and Discussion

3.1. Dough Rheology

The vast majority of food materials, including dough, exhibit rheological characteristics, which makes it impossible to classify their state as either solid or liquid. Such materials show both elastic and viscous properties [33]. Elastic properties are represented by the storage modulus (G′), which describes the energy temporarily stored in the sample that can be recovered, whereas viscous properties are described by the loss modulus (G″) that corresponds to the energy used for initiation of the flow that is irrevocably converted into shear heat [34]. Mechanical spectra of gluten-free doughs are presented in Figure S1.

Parameters of power law equations describing the visco-elastic properties of gluten-free dough enriched with cricket powder are presented in Table 1. The fit of the employed models to the experimental data was good, as indicated by the values of coefficient of determination (R^2), which exceeded 0.97. All investigated samples were characterized by the dominance of solid-like behavior indicated by the fact that the values of K′ were greater than K″. This is typical even for more sol-like materials, for example, starch paste [35]. Replacement of starch by cricket powder in amounts up to 6% resulted only in minor changes in rheological properties of the analyzed dough samples. The only relevant change observed was the decrease in complex viscosity (K*), which was a result of a decrease in both types of mechanical properties (K′ and K″). Similar values of n*, n′ and n″, determined for samples RD, DCP2 and DCP6, suggest that a minor decrease in viscosity was observed over a wide range of angular velocity values. This was the only change in the mechanical properties of the dough caused by replacement of starch by cricket powder in those samples. Further increase in the cricket powder to starch ratio in dough resulted in a significant decrease in viscosity along with an increase in all n equation parameters. This involved a stronger decrease in viscosity at higher shear forces compared to other dough samples.

Table 1. The viscoelastic properties of dough.

Sample	K *	n *	R^2	K′	n′	R^2	K″	n″	R^2
RD	51,550	0.347	0.994	55,460	0.135	0.965	12,750	0.121	0.984
DCP2	45,830	0.353	0.994	38,780	0.146	0.974	8877	0.125	0.989
DCP6	46,780	0.356	0.994	39,240	0.146	0.975	9570	0.123	0.983
DCP10	41,730	0.401	0.990	34,780	0.175	0.982	9054	0.146	0.979

RD—reference dough; DCP2, DCP6, DCP10—dough with 2, 6 and 10% substitution of starch with cricket powder.

3.2. Water Behavior of Dough and Crumb

Low-field NMR is a method used in food analysis since the 1990s. It allow one to measure the spin–lattice T_1 and spin–spin T_2 relaxation times, which characterize the molecular dynamics of water in a sample [30,31,36,37]. The parameters of molecular dynamics of water in the dough and crumb of bread were determined on the basis of the ^{1}H NMR tests and are presented in Table 2. The presence of two water fractions (bound and bulk) was found, which is a typical result for this type of material [38,39]. With the increase in the amount of starch substituted by CP, a significant decrease in the value of spin–net T_1 and both components of the spin–spin T_2 relaxation times was observed. This indicates that CP addition resulted in the decrease in the ratio of bulk-to-bound water fractions. The method of producing CP (roasting and grinding of insects) makes it hydrophobic instead of hydrophilic [40]. The results obtained therefore suggest that the introduction of CP leads to a greater availability of water for the biopolymers in the dough. This has influence on the viscoelastic properties of the dough—a network formed by starch and hydrocolloids (Table 1). The measurements of the relaxation time in the bread crumb show that after thermal processing the amount of bulk water fraction in relation to bound water fraction decreases with increasing amounts of CP additive. In the case of RB and BCP2, the value of the T_1 parameter was lower by approximately 15% after baking in comparison to RD and DCP2, respectively. The other two breads were characterized by a 20% decrease in the value of this relaxation time. There were no statistically significant changes in the value of the spin–spin relaxation time T_{22} for the crumb samples RB, BCP2 and BCP6 that would result from the presence of CP. At the same time, the comparison of the value of this parameter between dough and the respective crumbs shows a 3-fold decrease for the RB sample and a 2-fold decrease for the BCP10 sample. The fact that the T_{22} time was decreased in all the bread samples compared to the respective dough samples indicates that the baking process resulted in the removal of free water. The water available for biopolymers and hydrocolloids was largely retained in the structure. This can be evidenced by both a relatively small decrease in the T_1 value for crumbs and dough in individual samples and the absence of statistically significant changes in the value of both components of the spin–spin relaxation time.

There was no effect of the substitution of starch by CP on water activity at equilibrium condition (a_w) and limit water activity (a_p) of the crumb (Table 3). The transport rate (V_D) was lower in samples containing CP than in the reference bread. The transport rate limitation is the result of interactions between water and starch as well as between water and hydrocolloids. This confirms the previous suggestion based on the analysis of relaxation times that CP present in the bread crumb leads to increased availability of water to biopolymers. Also significantly lower was the rate of surface conduction (V_p) in samples that contained CP. Combined with the data obtained using low-field NMR, this result confirms the previously described changes in the molecular properties of water that are a consequence of the introduction of CP.

Table 2. Results of ^{1}H NMR study for dough and bread.

Sample	T_1 (ms)	T_{21} (ms)	T_{22} (ms)
RD	279.9 ± 3.1 ^{A}	5.24 ± 0.88 ^{A}	45.46 ± 0.76 ^{A}
DCP2	251.9 ± 2.3 ^{B}	3.16 ± 0.31 ^{B}	43.65 ± 0.56 ^{B}
DCP6	246.6 ± 0.9 ^{C}	2.25 ± 0.22 ^{C}	38.66 ± 0.30 ^{C}
DCP10	223.1 ± 0.9 ^{D}	2.17 ± 0.35 ^{C}	32.10 ± 0.19 ^{D}
RB	235.7 ± 1.5 ^{a}	1.39 ± 0.22 ^{c}	16.07 ± 0.41 ^{b}
BCP2	213.4 ± 0.6 ^{b}	2.43 ± 0.15 ^{a}	16.39 ± 0.31 ^{b}
BCP6	198.1 ± 0.8 ^{c}	2.52 ± 0.27 ^{a}	15.84 ± 0.32 ^{b}
BCP10	179.3 ± 0.6 ^{d}	2.83 ± 0.25 ^{a}	17.05 ± 0.84 ^{a}

Mean values denoted by different letters (uppercase for dough, lowercase for bread) differ statistically significantly ($p < 0.05$). NMR—Nuclear Magnetic Resonance; RD—reference dough; DCP2, DCP6, DCP10—dough with 2, 6 and 10% substitution of starch with cricket powder; RB—reference bread; BCP2, BCP6, BCP10—bread with 2, 6 and 10% substitution of starch with cricket powder.

Table 3. The results of water activity.

Sample	a_w (-)	a_p (-)	V_D (s^{-1})	V_p (s^{-1})
RB	0.925 ± 0.002 ^{a}	0.487 ± 0.013 ^{a}	0.024 ± 0.002 ^{a}	0.0030 ± 0.0001 ^{a}
BCP2	0.926 ± 0.003 ^{a}	0.503 ± 0.015 ^{a}	0.022 ± 0.002 ^{ab}	0.0026 ± 0.0001 ^{b}
BCP6	0.929 ± 0.006 ^{a}	0.641 ± 0.037 ^{a}	0.019 ± 0.004 ^{b}	0.0025 ± 0.0006 ^{b}
BCP10	0.910 ± 0.007 ^{a}	0.591 ± 0.016 ^{a}	0.019 ± 0.002 ^{b}	0.0018 ± 0.0002 ^{c}

Mean values denoted by different letters differ statistically significantly ($p < 0.05$). RB—reference bread; BCP2, BCP6, BCP10—bread with 2, 6 and 10% substitution of starch with cricket powder; a_p—limit water activity (intermediate); a_w—water activity at equilibrium condition (final); V_D—transport rate; V_p—rate of the surface conduction.

3.3. Crumb Texture

As commonly known, water content and activity have effects on the texture of bread. Texture profile analysis was conducted in order to evaluate these changes. The force required to squeeze the food between the teeth is a measure of the hardness, which is responsible for the perception of the freshness of food [41]. As stated in Table 4, the reference bread had the highest hardness and chewiness values. Moreover, the values of these parameters decreased with increasing amount of CP in the formula of the bread. Emulsifiers are used in baking technology to reduce crumb hardness [42]. The softening effect of CP could be connected with the emulsifying properties of cricket proteins. Similar effects on the structure of gluten-free crumbs were previously described by other authors who observed a decrease in crumb hardness after adding natural emulsifiers to dough [43–45]. Crumb cohesiveness, a parameter that describes the degree of deformation of the food structure before its breakage, significantly increased with the addition of CP. The increased consistency of the crumb in the case of CP-containing bread samples in comparison to the control sample is undoubtedly a desirable feature. GF breads usually have high susceptibility to fracture or crumbling [46]. Despite the fact that the springiness values did not differ significantly between the tested samples, CP incorporation significantly increased the ability of the crumb to return to its original state after compression, as evidenced by higher resilience values observed in all the enriched bread samples. This could be directly related to the high protein content in CP [18], which significantly affected the formation of the bread texture.

Table 4. Textural properties of breadcrumbs.

Sample	Hardness (N)	Springiness (%)	Cohesiveness (-)	Chewiness (-)	Resilience (-)
RB	37.21 ± 4.28 ^{a}	99.3 ± 1.5 ^{a}	0.556 ± 0.022 ^{b}	2238 ± 286 ^{a}	0.341 ± 0.028 ^{b}
BCP2	35.73 ± 1.53 ^{a}	99.3 ± 0.5 ^{a}	0.612 ± 0.068 ^{ab}	2096 ± 277 ^{ab}	0.400 ± 0.079 ^{a}
BCP6	25.08 ± 2.19 ^{b}	99.5 ± 2.2 ^{a}	0.645 ± 0.052 ^{a}	1726 ± 293 ^{b}	0.431 ± 0.035 ^{a}
BCP10	24.53 ± 1.79 ^{b}	99.9 ± 1.8 ^{a}	0.691 ± 0.062 ^{a}	1710 ± 77 ^{b}	0.443 ± 0.049 ^{a}

Mean values denoted by different letters differ statistically significantly ($p < 0.05$). RB—reference bread; BCP2, BCP6, BCP10—bread with 2, 6 and 10% substitution of starch with cricket powder.

4. Conclusions

While substitution of starch with CP may improve the nutritional value of gluten-free bread, it can also cause a number of changes in the properties of both the dough and the final product. Despite the fact that only small changes of macroscopic properties of dough were observed in these rheological analyses, the molecular-level analyses of water contained in the dough revealed that CP increases the availability of water for biopolymers, such as starch or hydrocolloids. This was probably an effect of binding the fat fraction. As a result, significant changes in water dynamics were also observed in the ready bread crumb samples. Moreover, it was shown that the introduction of CP leads to the reduction of hardness of the bread and improves its consistency. While the health-beneficial properties of edible insects are known, more research is needed in order to fully describe the health-promoting properties of bakery products supplemented with cricket powder.

Supplementary Materials: The following are available online at http://www.mdpi.com/2304-8158/8/7/240/s1, Figure S1. Mechanical spectra of gluten-free dough with cricket powder.

Author Contributions: Conceptualization, P.K.; Investigation, P.K., K.W., Ł.M., O.B., J.L., P.K. and H.B.; Methodology, P.K. and H.B.; Supervision, P.K.; Writing original draft, P.K., J.L., P.K. and H.B..

Funding: The authors received no financial support for the research, authorship, and/or publication of this article.

Acknowledgments: The authors thank Paulina Sarbak and Krzysztof Smarzyński (Students' Scientific Club of Food Technologists, Poznań University of Life Sciences, Poznań, Poland) for their help with the preparation of bread and the analyses.

Conflicts of Interest: On behalf of all authors, the corresponding author states that there is no conflict of interest.

References

1. Green, P.H.R.; Cellier, C. Celiac Disease. *N. Engl. J. Med.* **2007**, *357*, 1731–1743. [CrossRef]
2. Van Heel, D.A. Recent advances in coeliac disease. *Gut* **2006**, *55*, 1037–1046. [CrossRef] [PubMed]
3. Armstrong, M.J.; Robins, G.G.; Howdle, P.D. Recent advances in coeliac disease. *Curr. Opin. Gastroenterol.* **2009**, *25*, 100–109. [CrossRef] [PubMed]
4. Niewinski, M.M. Advances in Celiac Disease and Gluten-Free Diet. *J. Am. Diet. Assoc.* **2008**, *108*, 661–672. [CrossRef] [PubMed]
5. Green, P.H.R. The many faces of celiac disease: Clinical presentation of celiac disease in the adult population. *Gastroenterology* **2005**, *128*, S74–S78. [CrossRef] [PubMed]
6. Lebwohl, B.; Sanders, D.S.; Green, P.H.R. Coeliac disease. *Lancet* **2018**, *391*, 70–81. [CrossRef]
7. Lindfors, K.; Ciacci, C.; Kurppa, K.; Lundin, K.E.A.; Makharia, G.K.; Mearin, M.L.; Murray, J.A.; Verdu, E.F.; Kaukinen, K. Coeliac disease. *Nat. Rev. Dis. Prim.* **2019**, *5*, 3. [CrossRef] [PubMed]
8. Zannini, E.; Arendt, E.K. Low FODMAPs and gluten-free foods for irritable bowel syndrome treatment: Lights and shadows. *Food Res. Int.* **2018**, *110*, 33–41. [CrossRef] [PubMed]
9. Torbica, A.; Hadnadev, M.; Dapčević, T. Rheological, textural and sensory properties of gluten-free bread formulations based on rice and buckwheat flour. *Food Hydrocoll.* **2010**, *24*, 626–632. [CrossRef]
10. Gujral, H.S.; Rosell, C.M. Improvement of the breadmaking quality of rice flour by glucose oxidase. *Food Res. Int.* **2004**, *37*, 75–81. [CrossRef]
11. Lazaridou, A.; Duta, D.; Papageorgiou, M.; Belc, N.; Biliaderis, C.G. Effects of hydrocolloids on dough rheology and bread quality parameters in gluten-free formulations. *J. Food Eng.* **2007**, *79*, 1033–1047. [CrossRef]
12. Pacyński, M.; Wojtasiak, R.Z.; Mildner-Szkudlarz, S. Improving the aroma of gluten-free bread. *LWT-Food Sci. Technol.* **2015**, *63*, 706–713. [CrossRef]
13. Rybicka, I.; Doba, K.; Bińczak, O. Improving the sensory and nutritional value of gluten-free bread. *Int. J. Food Sci. Technol.* **2019**. [CrossRef]
14. Rybicka, I.; Gliszczyńska-Świgło, A. Minerals in grain gluten-free products. The content of calcium, potassium, magnesium, sodium, copper, iron, manganese, and zinc. *J. Food Compos. Anal.* **2017**, *59*, 61–67. [CrossRef]

15. Raheem, D.; Carrascosa, C.; Oluwole, O.B.; Nieuwland, M.; Saraiva, A.; Millán, R.; Raposo, A. Traditional consumption of and rearing edible insects in Africa, Asia and Europe. *Crit. Rev. Food Sci. Nutr.* **2018**, 1–20. [CrossRef] [PubMed]
16. Ghosh, S.; Jung, C.; Meyer-Rochow, V.B. What Governs Selection and Acceptance of Edible Insect Species. In *Edible Insects in Sustainable Food Systems*; Springer International Publishing: Cham, Switzerland, 2018; pp. 331–351.
17. Van Huis, A. Potential of Insects as Food and Feed in Assuring Food Security. *Annu. Rev. Entomol.* **2013**, *58*, 563–583. [CrossRef] [PubMed]
18. Montowska, M.; Kowalczewski, P.Ł.; Rybicka, I.; Fornal, E. Nutritional value, protein and peptide composition of edible cricket powders. *Food Chem.* **2019**, *289*, 130–138. [CrossRef] [PubMed]
19. Zielińska, E.; Baraniak, B.; Karaś, M.; Rybczyńska, K.; Jakubczyk, A. Selected species of edible insects as a source of nutrient composition. *Food Res. Int.* **2015**, *77*, 460–466. [CrossRef]
20. Kulma, M.; Kouřimská, L.; Plachý, V.; Božik, M.; Adámková, A.; Vrabec, V. Effect of sex on the nutritional value of house cricket, Acheta domestica L. *Food Chem.* **2019**, *272*, 267–272. [CrossRef]
21. Zielińska, E.; Baraniak, B.; Karaś, M. Identification of antioxidant and anti-inflammatory peptides obtained by simulated gastrointestinal digestion of three edible insects species (Gryllodes sigillatus, Tenebrio molitor, Schistocerca gragaria). *Int. J. Food Sci. Technol.* **2018**, *53*, 2542–2551. [CrossRef]
22. Zielińska, E.; Baraniak, B.; Karaś, M. Antioxidant and Anti-Inflammatory Activities of Hydrolysates and Peptide Fractions Obtained by Enzymatic Hydrolysis of Selected Heat-Treated Edible Insects. *Nutrients* **2017**, *9*, 970. [CrossRef] [PubMed]
23. Pauter, P.; Różańska, M.; Wiza, P.; Dworczak, S.; Grobelna, N.; Sarbak, P.; Kowalczewski, P.Ł. Effects of the replacement of wheat flour with cricket powder on the characteristics of muffins. *Acta Sci. Pol. Technol. Aliment.* **2018**, *17*, 227–233. [PubMed]
24. Smarzyński, K.; Sarbak, P.; Musiał, S.; Jeżowski, P.; Piątek, M.; Kowalczewski, P.Ł. Nutritional analysis and evaluation of the consumer acceptance of pork pâté enriched with cricket powder-preliminary study. *Open Agric.* **2019**, *4*, 159–163. [CrossRef]
25. Duda, A.; Adamczak, J.; Chełmińska, P.; Juszkiewicz, J.; Kowalczewski, P. Quality and Nutritional/Textural Properties of Durum Wheat Pasta Enriched with Cricket Powder. *Foods* **2019**, *8*, 46. [CrossRef] [PubMed]
26. Korus, J.; Juszczak, L.; Ziobro, R.; Witczak, M.; Grzelak, K.; Sójka, M. Defatted strawberry and blackcurrant seeds as functional ingredients of gluten-free bread. *J. Texture Stud.* **2012**, *43*, 29–39. [CrossRef]
27. Kowalczewski, P.; Różańska, M.; Makowska, A.; Jeżowski, P.; Kubiak, P. Production of wheat bread with spray-dried potato juice: Influence on dough and bread characteristics. *Food Sci. Technol. Int.* **2019**, *25*, 223–232. [CrossRef] [PubMed]
28. Baranowska, H.M.; Masewicz, Ł.; Kowalczewski, P.Ł.; Lewandowicz, G.; Piątek, M.; Kubiak, P. Water properties in pâtés enriched with potato juice. *Eur. Food Res. Technol.* **2018**, *244*, 387–393. [CrossRef]
29. Brosio, E.; Gianferri, R.R. An analytical tool in foods characterization and traceability. In *Basic NMR in Foods Characterization*; Research Signpost: Kerala, India, 2009; pp. 9–37.
30. Baranowska, H.M. Water Molecular Properties in Forcemeats and Finely Ground Sausages Containing Plant Fat. *Food Biophys.* **2011**, *6*, 133–137. [CrossRef]
31. Płowaś-Korus, I.; Masewicz, Ł.; Szwengiel, A.; Rachocki, A.; Baranowska, H.M.; Medycki, W. A novel method of recognizing liquefied honey. *Food Chem.* **2018**, *245*, 885–889. [CrossRef]
32. Masewicz, L.; Lewandowicz, J.; Le Thanh-Blicharz, J.; Kempka, M.; Baranowska, H.M. Diffusion of water in potato starch pastes. In Proceedings of the 12th International Conference on Polysaccharides-Glycoscience, Prague, Czech Republic, 19–21 October 2016; pp. 193–195.
33. Abang Zaid, D.N.; Chin, N.L.; Yusof, Y.A. A Review on Rheological Properties and Measurements of Dough and Gluten. *J. Appl. Sci.* **2010**, *10*, 2478–2490. [CrossRef]
34. Yan, H.; Yayuan, Z.; Ling, Z.; Zhengbiao, G. Study on physicochemical characteristics of waxy potato starch in comparison with other waxy starches. *Starch-Stärke* **2011**, *63*, 754–759. [CrossRef]
35. Lewandowicz, J.; Baranowska, H.M.; Szwengiel, A.; Le Thanh-Blicharz, J. Molecular structure vs. Functional properties of waxy and normal corn starch. In Proceedings of the 12th International Conference on Polysaccharides-Glycoscience, Prague, Czech Republic, 19–21 October 2016; pp. 53–57.

36. Makowska, A.; Baranowska, H.M.; Michniewicz, J.; Chudy, S.; Kowalczewski, P.Ł. Triticale extrudates—Changes of macrostructure, mechanical properties and molecular water dynamics during hydration. *J. Cereal Sci.* **2017**, *74*, 250–255. [CrossRef]
37. Piątek, M.; Baranowska, H.M.; Krzywdzińska-Bartkowiak, M. Microstructure and water molecular dynamics in meat after thawing. *Fleischwirtschaft* **2013**, *93*, 100–104.
38. van Nieuwenhuijzen, N.H.; Tromp, R.H.; Mitchell, J.R.; Primo-Martín, C.; Hamer, R.J.; van Vliet, T. Relations between sensorial crispness and molecular mobility of model bread crust and its main components as measured by PTA, DSC and NMR. *Food Res. Int.* **2010**, *43*, 342–349. [CrossRef]
39. Curti, E.; Bubici, S.; Carini, E.; Baroni, S.; Vittadini, E. Water molecular dynamics during bread staling by Nuclear Magnetic Resonance. *LWT-Food Sci. Technol.* **2011**, *44*, 854–859. [CrossRef]
40. Bassett, F. Comparison of Functional, Nutritional, and Sensory Properties of Spray-Dried and Oven-Dried Cricket (Acheta Domesticus) Powder. Master's Thesis, Brigham Young University, Provo, UT, USA, 2018.
41. Giannou, V.; Tzia, C. Frozen dough bread: Quality and textural behavior during prolonged storage—Prediction of final product characteristics. *J. Food Eng.* **2007**, *79*, 929–934. [CrossRef]
42. Gray, J.A.; Bemiller, J.N. Bread Staling: Molecular Basis and Control. *Compr. Rev. Food Sci. Food Saf.* **2003**, *2*, 1–21. [CrossRef]
43. Conte, P.; Del Caro, A.; Balestra, F.; Piga, A.; Fadda, C. Bee pollen as a functional ingredient in gluten-free bread: A physical-chemical, technological and sensory approach. *LWT* **2018**, *90*, 1–7. [CrossRef]
44. Alvarez-Jubete, L.; Auty, M.; Arendt, E.K.; Gallagher, E. Baking properties and microstructure of pseudocereal flours in gluten-free bread formulations. *Eur. Food Res. Technol.* **2010**, *230*, 437–445. [CrossRef]
45. Demirkesen, I.; Mert, B.; Sumnu, G.; Sahin, S. Rheological properties of gluten-free bread formulations. *J. Food Eng.* **2010**, *96*, 295–303. [CrossRef]
46. Onyango, C.; Mutungi, C.; Unbehend, G.; Lindhauer, M.G. Modification of gluten-free sorghum batter and bread using maize, potato, cassava or rice starch. *LWT-Food Sci. Technol.* **2011**, *44*, 681–686. [CrossRef]

© 2019 by the authors. Licensee MDPI, Basel, Switzerland. This article is an open access article distributed under the terms and conditions of the Creative Commons Attribution (CC BY) license (http://creativecommons.org/licenses/by/4.0/).

foods

Article

The Potential of Modulating the Reducing Sugar Released (and the Potential Glycemic Response) of Muffins Using a Combination of a Stevia Sweetener and Cocoa Powder †

Jingrong Gao [1,2,3,*], Xinbo Guo [1,4], Margaret A. Brennan [2,3], Susan L. Mason [2], Xin-An Zeng [1,4] and Charles S. Brennan [1,2,3,4,*]

1 School of Food Science and Engineering, South China University of Technology, Guangzhou 510640, China; guoxinbo@scut.edu.cn (X.G.); xazeng@scut.edu.cn (X.-A.Z.)
2 Department of Wine, Food and Molecular Biosciences, Lincoln University, Christchurch 7647, New Zealand; margaret.brennan@lincoln.ac.nz (M.A.B.); Sue.mason@lincoln.ac.nz (S.L.M.)
3 Riddet Research Institute, Palmerston North 4442, New Zealand
4 Overseas Expertise Introduction Center for Discipline Innovation of Food Nutrition and Human Health (111 Center), Guangzhou 510640, China
* Correspondence: gaojingrong@scut.edu.cn (J.G.); charles.brennan@lincoln.ac.nz (C.S.B.)

† The article is part of Ph.D. Thesis of Jingrong Gao.

Received: 16 October 2019; Accepted: 20 November 2019; Published: 5 December 2019

Abstract: Muffins are popular bakery products. However, they generally contain high amounts of sugar. The over-consumption of muffins may therefore result in a high calorie intake and could lead to increased health risks. For this reason, muffins were prepared substituting sucrose with two levels of a base of stevia (Stevianna®). In addition, cocoa powder and vanilla were added to the muffin formulation with and without Stevianna® to mask any potential off flavors. Results illustrate that muffins with 50% Stevianna® replacement of sucrose were similar to the control samples in terms of volume, density and texture. However, replacement of sugar with 100% Stevianna® resulted in reductions in height (from 41 to 28 mm), volume (from 63 to 51 mL), and increased firmness (by four-fold) compared to the control sample. Sugar replacement significantly reduced the in vitro predictive glycemic response of muffins (by up to 55% of the control sample). This work illustrates the importance of sugar in maintaining muffin structure as well as controlling the rate of glucose release during simulated digestions.

Keywords: muffin; in vitro starch digestibility; glycemic index; stevia; sugar replacement

1. Introduction

In recent years, consumers have gained an increasing awareness regarding the effect of dietary carbohydrates on the nutritional quality of foods. In particular, attention has been focused on the relationship between the various types of carbohydrate containing foods and the different postprandial glucose responses by these foods post ingestion [1–7]. The glycemic index (GI) is a physiological classification widely accepted for carbohydrate foods based on their ability to raise the concentration of glucose in the blood [7–9]. Bakery foods, muffins for example, are regarded as a high glycemic impact food due to the high concentration of sugar contained in the muffins. Previous research [10,11] has shown that the over-consumption of sucrose can lead to a number of metabolic complications including hyperinsulinemia, hyperglycemia, hypertension and insulin resistance, as well as being related to dyslipidemia and ectopic lipid deposition in healthy subjects with diabetes [12]. Indeed, high GI food products are quickly digested and their carbohydrate is rapidly absorbed, resulting in higher blood glucose levels [13]. On the contrary, the health benefits of the low GI products are thought to be derived from the slower the rate of carbohydrate absorption, consequently leading to a gradual rise in blood glucose level and better glycemic control [14].

The food industry has focused on reducing the calorific content of food to promote a healthier diet. Therefore, different natural sweeteners have been used in sugar-reduced or sugar-free products based on their multiple potential health benefits and functional properties, including maintaining sweetness and acceptable texture [15–18].

Steviol glycosides have been extracted and purified from the leaves of *Stevia rebaudiana* Bertoni, commonly known as stevia; they are naturally sweet-tasting, have good solubility in water, good temperature and pH stability [19–21] as well as having no calorific value [22], allowing them to be used as a sugar substitute or natural sweetener. Stevioside and rebaudioside A are the major glycoside constituents responsible for sweetness and are the most abundant glycosides in the *Stevia rebaudiana* Bertoni plant [23–25]. They are very useful as a food additive due to their relative sweetness being 250–300 times sweeter than table sugar [26].

Extracts from stevia have broad health-promoting properties for blood glucose and insulin levels in human studies [27]. Steviol glycosides are not hydrolyzed by human digestive enzymes of the mouth, stomach, and small intestine [28]. However, rebaudioside A and stevioside are hydrolyzed (in vitro and in vivo) to aglycone steviol by colon microflora through the successive removal of glucose units [29]. Chang et al. [27] reported that insulin sensitivity is increased due to stevia consumption in rodent models, and thus does not increase blood glucose and insulin levels [22]. Furthermore, previous work has found that a reduction in the predicted glycemic response was observed due to 50% or 100% replacement of sucrose with Stevianna® in muffins during in vitro digestion experiments [30]. Therefore, stevia has the potential to be a low-cost natural sweetener due to important pro-health properties, such as being non-calorific, non-fermentable and non-toxic as well as having a high-intensity sweetness [31], and it is also recommended as a treatment for diabetics and obese persons [23].

However, several studies have shown that the utilization of stevia as a sugar replacer in baking leads to a negative effect on appearance, compactness, moisture and texture of the bakery products structure [17,32,33]. These results have indicated that stevia is not acceptable to replace sucrose completely in bakery products as stevia exhibits high-intensity sweetness but does not possess the necessary bulking characteristics [34]. That is why Stevianna® (product code ST001 SE supplied by Stevianna® NZ) is used for our study, as it incorporates rebaudioside A (98% steviol glycoside; 1%) with erythritol (99%).

Erythritol is a four-carbon sugar alcohol or polyol with approximately 60% to 80% of the sweetness of sucrose [35]. It is not only a sweetener but also a bulking agent, and thus can be used as a sugar replacer in bakery products. Partial replacement of sucrose with erythritol had no negative influence on physical quality characteristics in a baked product [34,36]. In addition, previous studies reported that erythritol is useful as a non-glycemic and low-calorie sweetener that is safe for diabetics [37,38]. Erythritol has been demonstrated to have a small molecular size, thus it is rapidly absorbed by the small intestine and does not undergo systemic metabolism by the human body [37,39]. Some research has shown that the combination of a high-intensity sweetener with bulking agents or fibers in sugar-reduced formulations of food resulted in bakery products with acceptable physical quality [26,29,40,41].

None of these previous studies assessed a complex food sweetener to replace traditional sugar in bakery products. The aim of the study was to evaluate the replacement of sugar with Stevianna® (1 × sweetness of sucrose) and the addition of cocoa powder and/or vanilla to muffins for their physical properties and glycemic response, compared with a control muffin formulation with no added Stevianna®, cocoa powder, or vanilla.

2. Materials and Methods

2.1. Raw Materials

Wheat flour (Medal Premium baker flour, Champion, Auckland, New Zealand), white sugar (Chelsea, Auckland, New Zealand), baking powder (Edmonds, Christchurch, New Zealand), iodized table salt (Cerebos, Auckland, New Zealand), skim milk powder (Pams, Auckland, New Zealand), 100% cocoa powder (Cadbury, Dunedin, New Zealand), vanilla (Hansells, Sydney, Australia), canola oil (Pams, Auckland, New Zealand), and fresh eggs were purchased from a local supermarket and tap water was used. Muffins were prepared containing 0%, 50% and 100% Stevianna® (produce code ST001_SE; Stevianna®, Auckland, New Zealand) as a replacement for sucrose. Stevianna® utilizes Reb-A 98% steviol glycoside as the main sugar substitute along with erythritol.

2.2. Muffin Preparation

The muffin recipe was adapted from a previous study [30] and is given in Table 1. The Stevianna® was dissolved in the water and mixed with liquid whole egg and oil. After that, the dry ingredients were added into the liquid components and mixed for 5 min. The batter was poured into a paper baking case in a muffin pan. The muffins were baked for 18 min in a preheated Simpson Gemini Atlas series oven at 180 °C set to fan bake. Baked muffins were cooled at room temperature for 1 h, then packed in plastic resealable bags and stored in a refrigerator at 4 °C until physical analysis.

2.3. Muffin Height

The muffin product was taken out from the paper baking case, and the muffin height was measured with an electronic caliper (INSIZE) from the highest point of the muffin to the bottom of the muffin.

2.4. Moisture Content

A domestic kitchen food chopper (Zyliss®) was used to crush and homogenize the muffin (crust and crumb) of each formulation. Approximately 4 g was dried in an air oven at 105 °C for 16 h, until no further weight change.

The moisture content (MC) was calculated using the following equation:

$$MC\ (\%) = (W_{\text{before drying}} - W_{\text{after drying}}/W_{\text{before drying}}) \times 100 \quad (1)$$

where W denotes weight (g).

2.5. Muffin Volume

The volume of the muffins was measured by the rapeseed displacement method. Each muffin was placed in a plastic beaker of known volume (total volume, Vt), and the remaining space in the plastic beaker was then filled with rapeseed; the volume of the rapeseed required (Vs) was then determined by graduated cylinder. Muffin volume was calculated as the difference between the total volume and volume of rapeseed—the muffin volume = Vt − Vs [36].

2.6. Muffin Texture

A texture analyzer (TA.XT. Plus, Stable Microsystems, Surrey, UK) was used to measure the texture profile of muffins in terms of the firmness and springiness of the samples. The samples were compressed to a strain of 25% of the original height using a 75 mm cylindrical probe and a 50 kg load cell, and a test speed of 1.0 mm/s was used. Data was obtained from the Texture expert software (Stable Microsystems, Surrey, UK). Firmness and springiness values were calculated as the overall force of compression required and the resistance post compression.

Table 1. Formulas for muffins at two Stevianna levels, with or without cocoa powder and/or vanilla.

Formulation [a]	C	V	CP	CP + V	50S	50S + V	50S + CP	50S + CP + V	100S	100S + V	100S + CP	100S + CP + V
Ingredients	**Mass (g)**											
Wheat flour	138.4	138.4	115.3	115.3	138.4	138.4	115.3	115.3	138.4	138.4	115.3	115.3
Sugar	92.2	92.2	92.2	92.2	46.1	46.1	46.1	46.1	0	0	0	0
Baking powder	6.5	6.5	6.5	6.5	6.5	6.5	6.5	6.5	6.5	6.5	6.5	6.5
Salt	1.4	1.4	1.4	1.4	1.4	1.4	1.4	1.4	1.4	1.4	1.4	1.4
Skim milk powder	8.7	8.7	8.7	8.7	8.7	8.7	8.7	8.7	8.7	8.7	8.7	8.7
Oil	77.6	77.6	77.6	77.6	77.6	77.6	77.6	77.6	77.6	77.6	77.6	77.6
Liquid whole egg	34.6	34.6	34.6	34.6	34.6	34.6	34.6	34.6	34.6	34.6	34.6	34.6
Top water	97.6	97.6	97.6	97.6	97.6	97.6	97.6	97.6	97.6	97.6	97.6	97.6
Cocoa powder	0	0	23.1	23.1	0	0	23.1	23.1	0	0	23.1	23.1
Vanilla	0	3	0	3	0	3	0	3	0	3	0	3
Stevia	0	0	0	0	46.1	46.1	46.1	46.1	92.2	92.2	92.2	92.2

[a] Sample name of formulation: Control (C); Vanilla (V); Cocoa Powder (CP); Cocoa + Vanilla (CP + V); 50% Stevianna (50S); 50% Stevianna + Vanilla (50S + V); 50% Stevianna + Cocoa (50S + CP); 50% Stevianna + Cocoa + Vanilla (50S + CP + V); 100% Stevianna (100S); 100% Stevianna + Vanilla (100S + V); 100% Stevianna + Cocoa (100S + CP); 100% Stevianna + Cococa + Vanilla (100S + CP + V).

2.7. Muffin Total Starch

Total starch analysis was carried out according to the official American Association of Cereal Chemists method 76.13 [42], using Megazyme (Bray, Dublin, Ireland) total starch kit.

2.8. In Vitro Predictive Glycemic Response Digestion Analysis

The procedure used for the determination of potential glycemic response is the same as that reported previously by [30]. This procedure measures the breakdown of carbohydrates to sugars by the action of amylase enzymes added to the baked muffin. Whole muffins were chopped with a domestic kitchen food chopper (Zyliss®) to stimulate particle size reduction which occurs during natural mastication for at least one minute of steady chopping until a fine crumb was achieved. A 3.5 g sample was used to determine the predictive glycemic response.

Triplicate samples of product (approximate 1 g of cooked muffin) were each placed into the 60 mL plastic pots and 30 mL of distilled water added, and duplicate blank samples. These pots were inserted to a pre-heated 15 place magnetic heated stirring block (IKAMAG® RT15, IKA®-WERKE Gmblt & Co., Staufen, Germany) preheated to 37 °C, on each pot one magnetic stirrer, followed by 0.8 mL of 1 M aqueous HCl. Then, 1 mL of a 10% pepsin (Acros Organics, New Jersey, NJ, USA CAS: 901-75-6) solution in 0.05 M HCl was added in order to replicate gastric digestion. The sample was incubated at 37 °C for 30 min with slow constant stirring (130 rpm) to simulate gastric digestion conditions. In vitro stomach digestion was halted by the addition of 2 mL $NaHCO_3$. Small intestine digestion was mimicked by the addition of 5 mL 0.1 M Na maleate buffer pH 6. An aliquot (1 mL) was withdrawn (Time 0) and added to 4 mL absolute ethanol to stop any further enzyme reaction. A 0.1 mL dose of amyloglucosidase (A.niger, Megazyme, E-AMGDF; 3260 U/mL) was added to prevent end-product inhibition of pancreatic amylase. A 5 mL 2.5% pancreatin (EC: 232-468-9, CAS: 8049-47-6, activity: 42362 FIP-U/g, Applichem GmbH, Darmstadt, Germany) in 0.1 M Na maleate buffer pH 6 followed by the volume being made to 53 mL with continued stirring and heat maintained at 37 °C for 120 min. Triplicate 1 mL aliquots were withdrawn at 0, 20, 60, 120 min and added to 4 mL absolute ethanol. Reducing sugar content was analyzed by dinitrosalicyclic (DNS) colorimetry, and the area under the curve (AUC) was calculated by dividing the graph into trapezoids as described elsewhere [30]. The reducing sugar content was regarded as an indicator for the predictive glycemic response.

2.9. Statistical Analyses

All analyses were conducted in triplicate. Analysis of variance (one-way ANOVA) was performed on the data, and Tukey's comparison test ($p < 0.05$) was used to determine the significance. These analyses were performed using Minitab (Minitab Pty Ltd., Sydney, Australia).

3. Results and Discussion

3.1. Moisture Content

Table 2 shows that the moisture content of muffin samples ranged from 19% to 27%. The moisture content of the muffin samples produced was higher when cocoa powder or/and vanilla was used. In addition, Figure 1 shows that moisture content values increased significantly ($p < 0.05$) when sucrose was replaced by Stevianna®—in particular the moisture content of 100% Stevianna® samples were higher than the full-sucrose muffin samples. Sucrose plays an important role in water retention that results in reduced moisture loss during the baking of the muffins [43]. However, the moisture content increased when sucrose was replaced because the Stevianna® acted as a humectant and prevented water from escaping during baking. Research using other types of sugar replacers has shown similar results. Martínez-Cervera et al. [44] used erythritol in muffins for its water retention properties. Ghosh and Sudha [45] showed that the use of the polyol sorbitol was reflected in a significantly higher moisture content ($p < 0.05$). Due to the high water-binding capacity of formulations with carbohydrate-based sugar replacers, a greater amount of water is required in cereal products.

Table 2. Effect of Stevianna on texture profile analysis and total starch in muffins with or without cocoa powder and/or vanilla.

Product	Firmness (g)	Springiness (%)	Total Starch (%)
C	746.06 ± 44.10 ^{b}	51.29 ± 0.44 ^{ab}	26.83 ± 1.92 ^{abc}
V	763.51 ± 51.48 ^{b}	51.66 ± 0.09 ^{a}	27.93 ± 0.42 ^{ab}
CP	680.99 ± 30.33 ^{b}	49.26 ± 0.54 ^{ab}	26.14 ± 0.60 ^{abcd}
CP + V	662.97 ± 68.46 ^{b}	49.99 ± 0.43 ^{ab}	24.43 ± 1.06 ^{bcde}
50S	906.07 ± 111.09 ^{b}	51.51 ± 0.62 ^{ab}	28.50 ±0.85 ^{a}
50S + V	1102.18 ± 102.10 ^{b}	51.49 ± 0.78 ^{a}	29.03 ± 0.36 ^{a}
50S + CP	987.03 ± 68.00 ^{b}	48.67 ± 0.52 ^{a}	22.72 ± 0.39 ^{de}
50S + CP + V	890.78 ± 76.18 ^{b}	49.59 ± 0.54 ^{b}	23.40 ± 0.09 ^{cde}
100S	4512.78 ± 399.65 ^{a}	45.07 ± 0.71 ^{c}	26.60 ± 0.94 ^{abc}
100S + V	4419.70 ± 409.69 ^{a}	45.44 ± 0.56 ^{c}	29.09 ± 2.56 ^{a}
100S + CP	3868.00 ± 300.87 ^{a}	44.74 ± 1.12 ^{c}	22.62 ± 1.42 ^{e}
100S + CP + V	3839.94 ± 522.34 ^{a}	43.11 ± 1.36 ^{c}	26.17 ± 1.14 ^{abcd}

Control (C); Vanilla (V); Cocoa Powder (CP); Cocoa+Vanilla (CP + V); 50% Stevianna (50S); 50% Stevianna + Vanilla (50S + V); 50% Stevianna + Cocoa (50S + CP); 50% Stevianna + Cocoa + Vanilla (50S + CP + V); 100% Stevianna (100S); 100% Stevianna + Vanilla (100S + V); 100% Stevianna + Cocoa (100S + CP); 100% Stevianna + Cococa + Vanilla (100S + CP + V). All measurements are the mean values ± SD of triplicate determinations. Means in the same column with different letters are significantly different ($p < 0.05$).

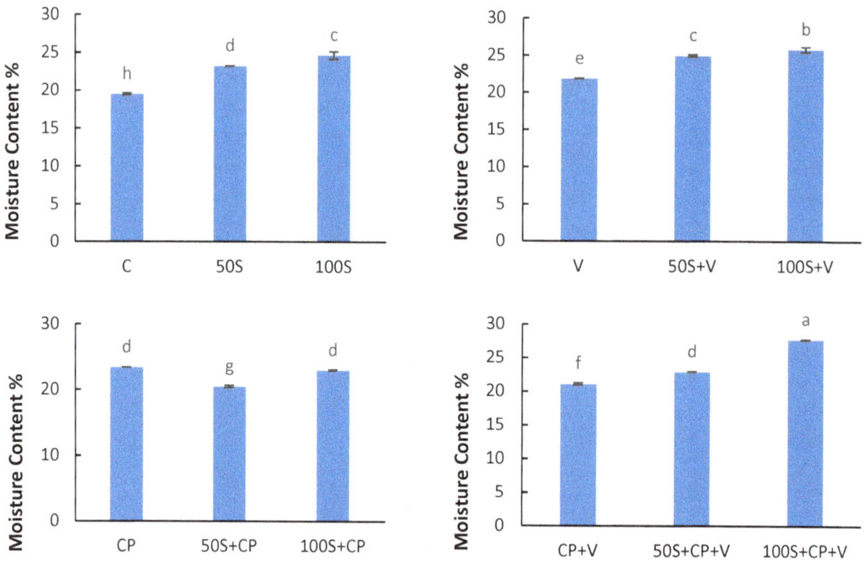

Figure 1. Moisture content for muffins of formulation made from two levels of Stevianna without/with cocoa powder and/or vanilla. Control (C); Vanilla (V); Cocoa Powder (CP); Cocoa + Vanilla (CP + V); 50% Stevianna (50S); 50% Stevianna + Vanilla (50S + V); 50% Stevianna + Cocoa (50S + CP); 50% Stevianna + Cocoa + Vanilla (50S + CP + V); 100% Stevianna (100S); 100% Stevianna + Vanilla (100S + V); 100% Stevianna + Cocoa (100S + CP); 100% Stevianna + Cococa + Vanilla (100S + CP + V). Values with different letters are significantly different to one another $p < 0.05$.

Moisture content in bakery products is an important factor as it has a direct impact on the texture attributes and a strong correlation has been found between moisture content and firmness [46]. As can be seen from the Table 2, muffin firmness increased as moisture content increased. As reported by Rößle et al. [47], this must be related to the replacement of the sugar by Stevianna®, affecting the formation of muffin structure.

3.2. The Impact of Sugar Replacement on Product Physico-Chemical Characteristics

The height of the muffins prepared with the different levels of Stevianna® with/without cocoa powder and/or vanilla is shown in Figure 2. The full-sucrose muffin was significantly higher ($p < 0.05$) than the muffins that were prepared using Stevianna®. The lowest height was found in the 100% Stevianna® muffin samples. The full-sucrose muffin with cocoa powder and/or vanilla group had a greater height than the control and other samples (Figure 2). These results indicate that the decrease in muffin height was associated with an absence of interconnectivity of a more compact structure and with a low number of air cells for levels of sucrose replacement higher than 50% (Figure 3).

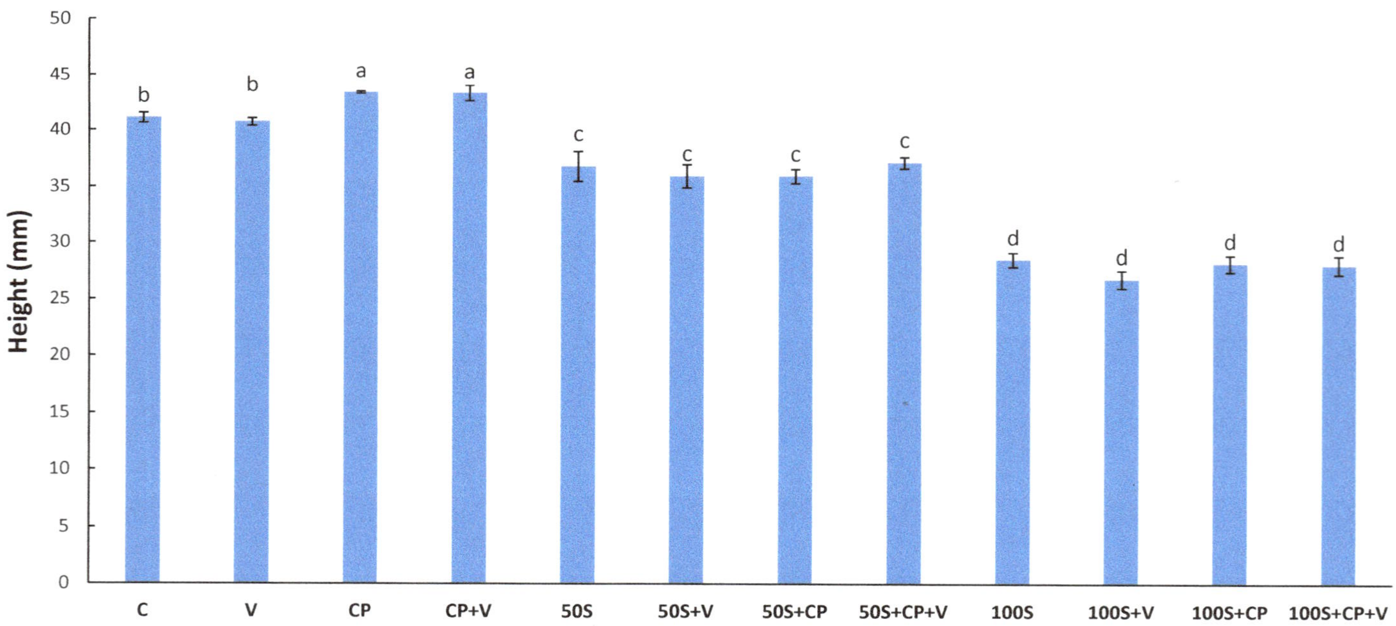

Figure 2. Effect of Stevianna without/with cocoa powder and/or vanilla on the height of muffin. Control (C); Vanilla (V); Cocoa Powder (CP); Cocoa + Vanilla (CP + V); 50% Stevianna (50S); 50% Stevianna + Vanilla (50S + V); 50% Stevianna + Cocoa (50S + CP); 50% Stevianna + Cocoa + Vanilla (50S + CP + V); 100% Stevianna (100S); 100% Stevianna + Vanilla (100S + V); 100% Stevianna + Cocoa (100S + CP); 100% Stevianna + Cococa + Vanilla (100S + CP + V). Values with different letters are significantly different to one another $p < 0.05$.

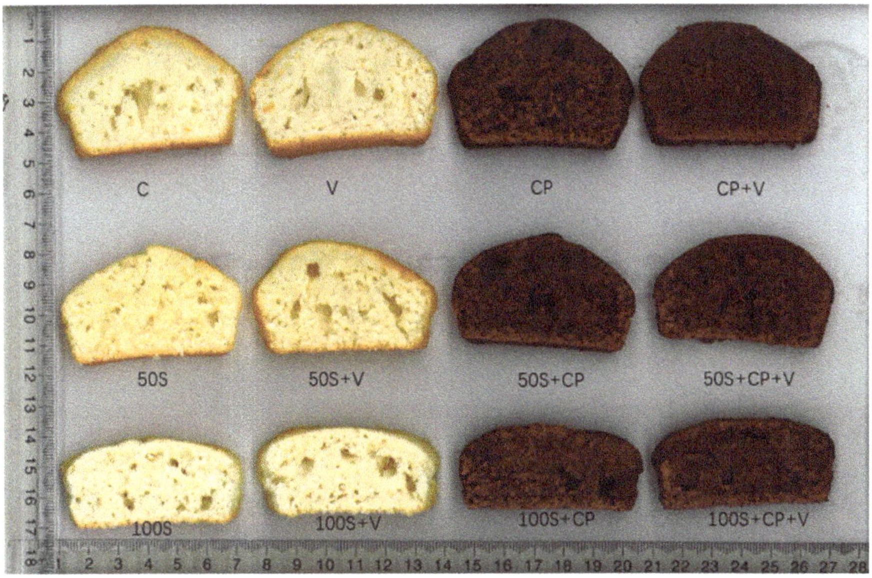

Figure 3. Effect of two levels of Stevianna with/without cocoa powder and/or vanilla in muffins: Control (C); Vanilla (V); Cocoa Powder (CP); Cocoa + Vanilla (CP + V); 50% Stevianna (50S); 50% Stevianna + Vanilla (50S + V); 50% Stevianna + Cocoa (50S + CP); 50% Stevianna + Cocoa + Vanilla (50S + CP + V); 100% Stevianna (100S); 100% Stevianna + Vanilla (100S + V); 100% Stevianna + Cocoa (100S + CP); 100% Stevianna + Cococa + Vanilla (100S + CP + V).

Photographs of vertical cross-sections of the different muffin formulations are shown in Figure 3. As the Stevianna® content increased, in the formulations, the air bubbles became smaller and the air channels gradually diminished. This could be due to the fact that muffins with a full sucrose content gained an increased number of air bubbles during the beating of the batter, and these air bubbles are then expanded by carbon dioxide and water vapor pressure generated during baking, resulting in the formation of air channels, which influence the texture of the finished muffin product. The lack of air channels as the sucrose was replaced may also be associated with earlier thermosetting of the batter during the heating process in the oven, therefore, not allowing enough time for bubble expansion and formation of air channels [43,44]. Martínez-Cervera et al. [44] also found that the number of small air bubbles increased, air channels diminished, and circular bubbles increased with an increase in sucrose replacement by polydextrose and sucralose in a muffin product.

The volume of the muffin is an important indicator of air bubble expansion during baking and consequently also of the porous structure of the product. The volumes of muffins prepared with different levels of Stevianna® with/without and/or vanilla along with the control muffin are presented in Figure 4A. The samples with 100% Stevianna® muffin group had significantly lower volumes ($p < 0.05$) compared to those of the full-sucrose muffin products. Muffin density appeared to be negatively correlated with muffin volume (Figure 4B). The density of the muffins was calculated from mass and volume after baking. Table 2 illustrates that when sugar was completely substituted with Stevianna®, there was a significant increase ($p < 0.05$) in muffin density. Additionally, product quality characteristics such as springiness and firmness were greatly affected (Table 2). These results indicate that an increase in the level of Stevianna® had an adverse effect on volume, density and texture of the muffin. Manisha et al. [26] also reported that replacement of sucrose with 100% stevioside and liquid sorbitol caused a significant deterioration in quality which decreased volume and resulted in a firmer texture in cake properties.

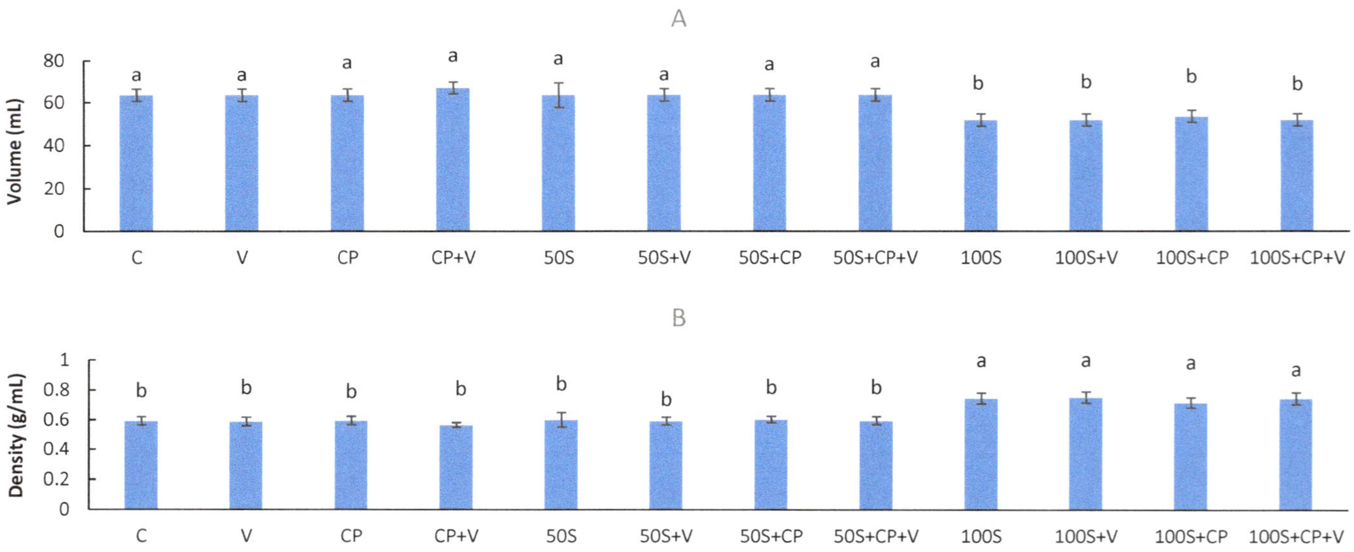

Figure 4. Volume (**A**) and density (**B**) values for muffins containing two levels of Stevianna as sugar replacer with or without cocoa powder and vanilla. Control (C); Vanilla (V); Cocoa Powder (CP); Cocoa + Vanilla (CP + V); 50% Stevianna (50S); 50% Stevianna + Vanilla (50S + V); 50% Stevianna + Cocoa (50S + CP); 50% Stevianna + Cocoa + Vanilla (50S + CP + V); 100% Stevianna (100S); 100% Stevianna + Vanilla (100S + V); 100% Stevianna + Cocoa (100S + CP); 100% Stevianna + Cococa + Vanilla (100S + CP + V). Values with different letters are significantly different to one another $p < 0.05$.

A function of sugar during cake baking is that it delays starch gelatinization, thus contributing to the aeration of the batter and the optimum quality of sugar will affect formation of the cake structure and improve crumb texture and tenderness [26]. The decrease in sugar-free muffin expansion is the result of less air bubble incorporation and reduced air holding capacity during baking [48]. In addition, starch gelatinization temperature seems to contribute to volume development due to different interactions between the Stevianna® and starch and proteins of the batter, and these interactions affect starch gelatinization and protein denaturation temperatures. These results are in agreement with Ronda et al. [49]'s findings which showed that a decrease in starch gelatinization and protein denaturation temperatures in sorbitol cakes is expected to cause a premature thermosetting of protein or starch matrix—this process will start at the crust due to direct contact with the heating medium. Therefore, this lowers the heat transfer rate, and produces a vapor pressure build-up, resulting in inadequate expansion of individual bubbles. Additionally, Ronda et al. [49] found that high-fructose corn syrup (HFCS) mainly contributed to the early gelatinization of starch during the baking process and restricted the volume of baked products compared to sucrose.

However, the 50% Stevianna® used had no significant effect on the volume and density of muffin compared to the full-sucrose muffin samples (Figure 4). These results suggest that muffin samples containing half the amount of Stevianna® have a similar ability, compared with muffins with full sucrose, to retain air. These results are consistent with those of Lin et al. [38], who found no significant differences among the volume estimates for 50% erythritol cakes. Furthermore, the addition of the 50% Stevianna® in muffin samples exhibited a texture close to that of the full-sucrose muffin samples (Table 2), which conferred an appearance of firmness and springiness. The results were consistent with previous research [30].

3.3. The Impact of Sugar Replacement on the In Vitro Predictive Glycemic Response

The total starch of modified muffins was measured and compared with the control sample (Table 2). Compared to the control muffin, 50% or 100% sucrose replacement with Stevianna® with added cocoa powder samples had significantly lower amounts of total starch. Similar levels of total starch were observed in control and full-sucrose muffin samples—50% and 100% Stevianna® with/without cocoa powder and/or vanilla muffin samples. Thus, the presence of cocoa powder with Stevianna® in muffin had a significant effect on total starch contents.

The effects of Stevianna® on in vitro starch digestion in muffin and chocolate muffin products were investigated by measuring the glucose released during starch digestion. Figure 5 shows the reducing sugars curves of two levels of Stevianna® with/without cocoa powder and/or vanilla muffin samples that were compared with full-sucrose with/without cocoa powder and/or vanilla samples, respectively. These two levels of Stevianna® used in this study were found to decrease reducing sugars released by digestive enzymes, compared with the full-sucrose muffin samples. The rate and extent of reducing sugars released were the highest in the control muffin, followed by 50% Stevianna® with/without cocoa powder and/or vanilla muffin products, and 100% Stevianna® with/without cocoa powder and/or vanilla muffins (Figure 5). In particular, muffins with Stevianna® showed a significant decrease in terms of reducing sugars released throughout the 120 min starch digestion process.

The total area under the hydrolysis curve (AUC) relates the total glucose release to the digestion time of 120 min. The concentration of the Stevianna® had a significant effect on the AUC values ($p < 0.05$), which demonstrated that the replacement of sucrose with 100% Stevianna® resulted in the lowest AUC value of muffin samples in a dose response (Figure 6). It is of interest that the additions of vanilla and/or cocoa powder with muffin production did not lead to a significant reduction of in vitro digestion values compared to the full-sucrose—50% Stevianna®, and 100% Stevianna® samples, respectively. These results are consistent with the previous report by Gao et al. [30].

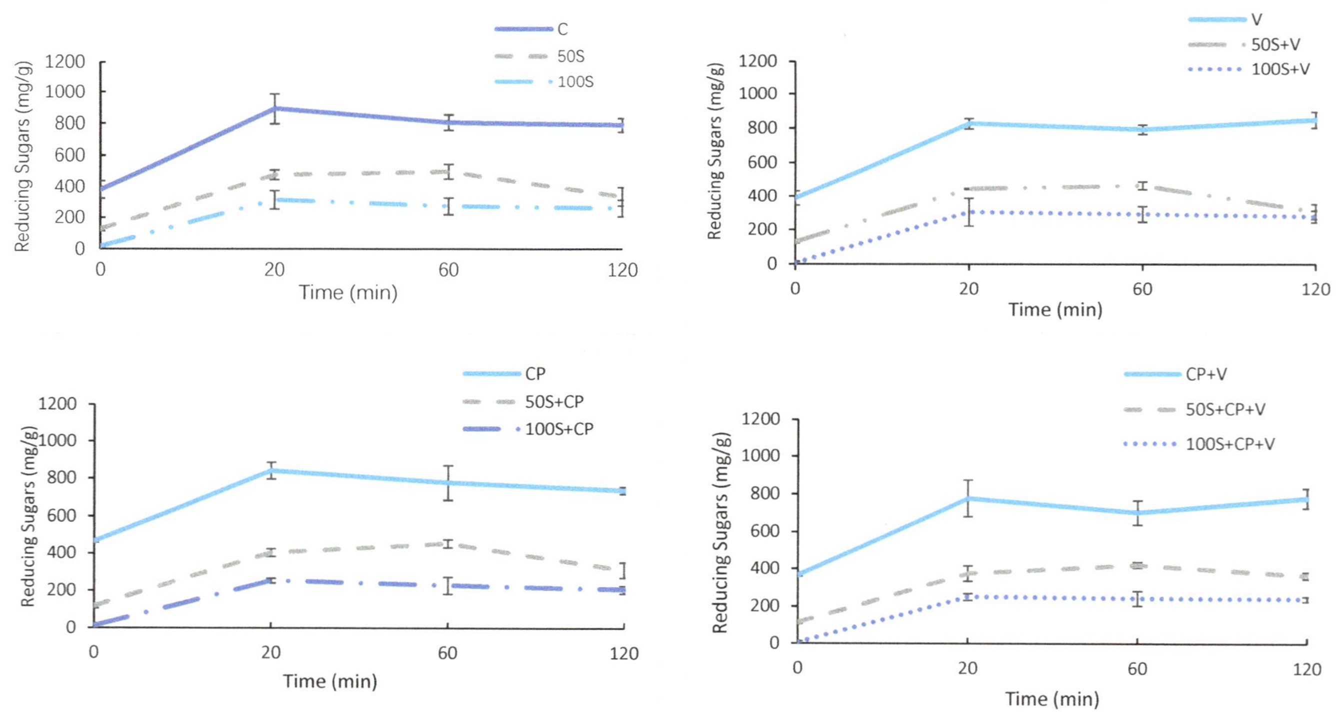

Figure 5. Amount of reducing sugars released per g of food material during in vitro digestion. Control (C); Vanilla (V); Cocoa Powder (CP); Cocoa + Vanilla (CP + V); 50% Stevianna (50S); 50% Stevianna + Vnilla (50S + V); 50% Stevianna + Cocoa (50S + CP); 50% Stevianna + Cocoa + Vanilla (50S + CP + V); 100% Stevianna (100S); 100% Stevianna + Vanilla (100S + V); 100% Stevianna + Cocoa (100S + CP); 100% Stevianna + Cococa + Vanilla (100S + CP + V).

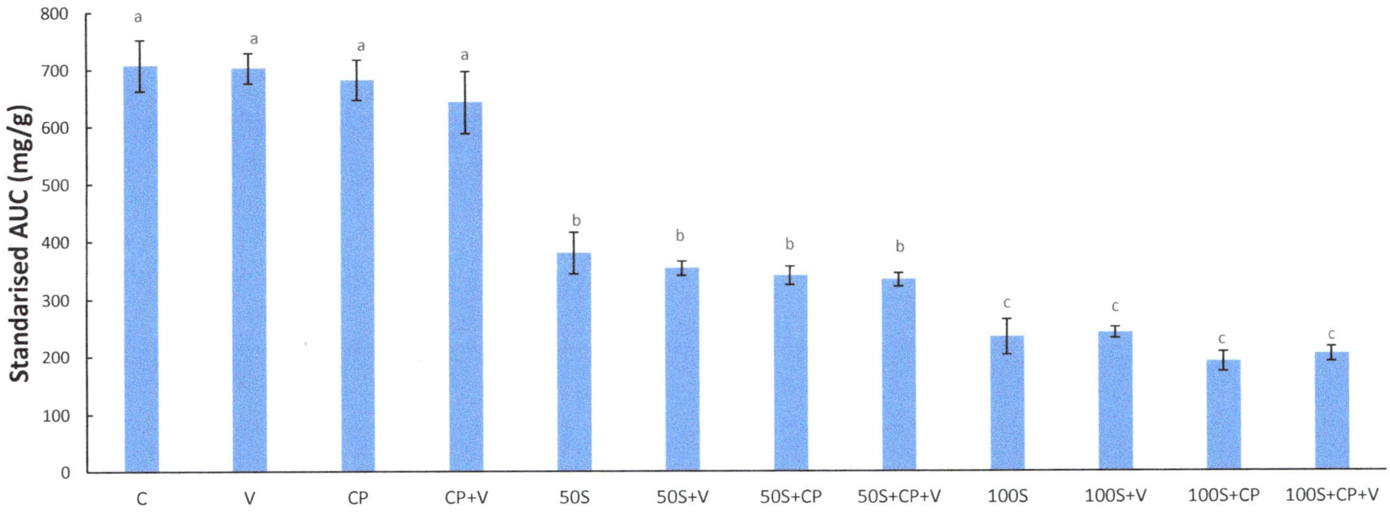

Figure 6. Values for area under the curve (AUC) comparing the control and other low-sugar muffins made with two levels of Stevianna with/without cocoa powder and/or vanilla. Control (C); Vanilla (V); Cocoa Powder (CP); Cocoa + Vanilla (CP + V); 50% Stevianna (50S); 50% Stevianna + Vanilla (50S + V); 50% Stevianna + Cocoa (50S + CP); 50% Stevianna + Cocoa + Vanilla (50S + CP + V); 100% Stevianna (100S); 100% Stevianna + Vanilla (100S + V); 100% Stevianna + Cocoa (100S + CP); 100% Stevianna + Cococa + Vanilla (100S + CP + V). Values with different letters are significantly different to one another $p < 0.05$.

This study did not focus on the impact of sweeteners on in vitro starch digestion analysis of bakery products. However, several research projects have been designed to test the effects of the stevia or erythritol on postprandial glucose and insulin levels in vivo and in vitro digestion methods as compared to sucrose [50,51].

The breakdown or disruption of starch granules that results from salivary amylase causes a greater susceptibility of the granule to further enzyme degradation. This process will lead to more readily digestible starch, and hence create a higher blood glucose response [52]. The level of postprandial blood glucose is a major factor in predicting the profile of insulin resistance. Alizadeh et al. [50] found that there were differing effects on postprandial blood insulin levels that were dependent on the type and amount of sweetener consumed. The effect of the consumption of beverages containing stevia has been tested by measuring the in vivo glycemic impact [53], and it was found that postprandial glucose and insulin levels were significantly reduced in the stevia beverages compared to the sucrose beverages. These effects on postprandial glucose levels are mainly due to the lack of calories and carbohydrate content of Stevianna®, and thus there are no reducing sugars released. A similar trend has been observed in that the postprandial insulin levels were reduced in stevia ice cream samples compared to full-sucrose ice cream samples [50], and this is most likely due to the functional properties of stevia that results in no contribution to the available carbohydrate and glycemic response in food products. In addition, Roberts and Renwick [54] illustrated that steviol glycosides are not readily absorbed by the upper small intestine when it is administered orally to normal rat or human subjects. There are no human digestive enzymes present in the small intestine to hydrolyze the β-glycosidic linkages, resulting in limited small intestine digestion.

Lin et al. [36] illustrated that 0%–100% sugar replacement with erythritol in cookies decreased the carbohydrate contents by in vivo digestion. Since the calorie value of erythritol is approximately 0.4 kcal/g [39], it provides no energy to the body and thus it is not systemically metabolized nor fermented in the colon [37]. It has been suggested that the consumption of erythritol does not raise postprandial glycemic and insulin levels by oral ingestion in healthy human subjects [28]. In a previous study [39], more than 90% of erythritol is rapidly absorbed by the small intestine when eaten and is excreted unchanged in the urine.

The Stevianna® used in our study was composed of rebaudioside A (stevia) and erythritol and, therefore, the observations made are consistent with those made by the above studies. Our experiment results showed that under in vitro conditions a lower reducing sugar liberation took place when sucrose was replaced by Stevianna® in muffins, and consequently this can be beneficial to as it will decrease the postprandial blood glucose. Additionally, it is probable that the intake of these muffins decreases the rate of intestine absorption of glucose and delays gastric emptying.

4. Conclusions

The stevia-containing product, Stevianna®, has been shown to be a suitable sucrose replacement for a low-sucrose formulation of muffins. The results showed that 50% sugar replacement with Stevianna® had similar physical quality characteristics in terms of volume, density and texture to a control muffin. However, when the sugar was replaced by 100% Stevianna®, the muffin quality showed a reduction in volume, an increase in textural firmness and a correspondingly high density of the product when compared to the control muffin samples. Furthermore, Stevianna® was able to simulate sucrose functionality in muffins, producing an increase in moisture content in comparison with the full-sucrose muffins. The negative effect of Stevianna® on muffin properties can be associated with the fact that as the Stevianna® level was raised, it led to a reduction of air bubble expansion during the heating process (possibly due to the weakening of the starch–protein–sugar interface of the muffin, allowing for greater structural collapse) and thus a corresponding reduction in height. This research illustrates that Stevianna® is a major factor impacting on the physical characteristics of muffins. The addition of cocoa powder and/or vanilla did not affect the quality of muffins significantly.

In relation to the nutritional quality of the muffin products, the effect of Stevianna® inclusion on the predicted glycemic impact as determined by in vitro digestion illustrated the role of sugar in elevating the glycemic response during digestion. The replacement of sugar with increasing levels of Stevianna® was found to significantly decrease the potential glycemic response values, and this is most likely to be attributed to the fact that Stevianna® was not degraded into glucose units and acted as an inert filler within the muffin samples. Therefore the inclusion of cocoa powder and/or vanilla powder did not have a significant change to the predicted glycemic response values of the muffins.

The breakdown or disruption of starch granules that results from salivary amylase causes a greater susceptibility of the granule to further enzyme degradation. This process will lead to more readily digestible starch, and hence create a higher blood glucose response [52]. The level of postprandial blood glucose is a major factor in predicting the profile of insulin resistance. Alizadeh et al. [50] found that there were differing effects on postprandial blood insulin levels that were dependent on the type and amount of sweetener consumed. The effect of the consumption of beverages containing stevia has been tested by measuring the in vivo glycemic impact [53], and it was found that postprandial glucose and insulin levels were significantly reduced in the stevia beverages compared to the sucrose beverages. These effects on postprandial glucose levels are mainly due to the lack of calories and carbohydrate content of Stevianna®, thus there are no reducing sugars released. A similar trend has been observed in that the postprandial insulin levels were reduced in stevia ice cream samples compared to full-sucrose ice cream samples [50], and this is most likely due to the functional properties of stevia that results in no contribution to the available carbohydrate and glycemic response in food products. In addition, Roberts and Renwick [54] illustrated that steviol glycosides are not readily absorbed by the upper small intestine when it is administered orally to normal rat or human subjects. There are no human digestive enzymes present in the small intestine to hydrolyze the β-glycosidic linkages, resulting in limited small intestine digestion.

Lin et al. [36] illustrated that 0%–100% sugar replacement with erythritol in cookies decreased the carbohydrate contents by in vivo digestion. Since the calorie value of erythritol is approximately 0.4 kcal/g [39], it provides no energy to the body and thus it is not systemically metabolized nor fermented in the colon [37]. It has been suggested that the consumption of erythritol does not raise postprandial glycemic and insulin levels by oral ingestion in healthy human subjects [28]. In a previous study [39], more than 90% of erythritol is rapidly absorbed by the small intestine when eaten and is excreted unchanged in the urine.

Finally, it can be seen that a partial replacement of Stevianna® for sucrose with/without cocoa powder and/or vanilla in muffins gave a product with quality characteristics close to that of the full-sucrose muffin sample. At the same time, the reduction in potential glycemic response values was greater than would have been expected with 50% sucrose reduction and consequently providing a quality muffin that produces a lowered postprandial response with the potential associated health benefits.

Author Contributions: J.G., M.A.B., C.S.B., X.G. and X.-A.Z. conceived and designed the experiments; J.G. and X.G. performed the experiments; J.G., M.A.B., S.L.M. and C.S.B. analyzed the data; J.G., C.S.B. and M.A.B. were responsible for writing the manuscript.

Funding: The research was supported by Lincoln University postgraduate funding. This research was also supported by the 111 Project (B17018) as well as S&T projects of Guangdong Province (2015A030312001 and 2013B020203001).

Conflicts of Interest: The authors declare no conflict of interest.

References

1. Giuberti, G.; Gallo, A. Reducing the glycaemic index and increasing the slowly digestible starch content in gluten-free cereal-based foods: A review. *Int. J. Food Sci. Technol.* **2018**, *53*, 50–60. [CrossRef]
2. Sopade, P.A. Cereal processing and glycaemic response. *Int. J. Food Sci. Technol.* **2017**, *52*, 22–37. [CrossRef]

3. Connolly, A.; O'Keeffe, M.B.; Nongonierma, A.B.; Piggott, C.O.; FitzGerald, R.J. Isolation of peptides from a novel brewers spent grain protein isolate with potential to modulate glycaemic response. *Int. J. Food Sci. Technol.* **2017**, *52*, 146–153. [CrossRef]
4. Brennan, C.S. Dietary fibre, glycaemic response, and diabetes. *Mol. Nutr. Food Res.* **2005**, *49*, 560–570. [CrossRef]
5. Jenkins, D.J.; Wolever, T.M.; Taylor, R.H.; Barker, H.M.; Fielden, H.; Jenkins, A.L. Effect of guar crispbread with cereal products and leguminous seeds on blood glucose concentrations of diabetics. *Br. Med. J.* **1980**, *281*, 1248–1250. [CrossRef]
6. Monro, J.A. Faecal bulking efficacy of Australasian breakfast cereals. *Asia Pac. J. Clin. Nutr.* **2002**, *11*, 176–185. [CrossRef]
7. Monro, J.A.; Shaw, M. Glycemic impact, glycemic glucose equivalents, glycemic index, and glycemic load: Definitions, distinctions, and implications. *Am. J. Clin. Nutr.* **2008**, *87*, 237S–243S. [CrossRef]
8. Gao, J.; Wang, Y.; Dong, Z.; Zhou, W. Structural and mechanical characteristics of bread and their impact on oral processing: A review. *Int. J. Food Sci. Technol.* **2018**, *53*, 858–872. [CrossRef]
9. Klunklin, W.; Savage, G. Physicochemical, antioxidant properties and in vitro digestibility of wheat–purple rice flour mixtures. *Int. J. Food Sci. Technol.* **2018**, *53*, 1962–1971. [CrossRef]
10. Barros, C.M.M.R.; Lessa, R.Q.; Grechi, M.P.; Mouço, T.L.M.; Souza, M.; das, G.C.; Wiernsperger, N.; Bouskela, E. Substitution of drinking water by fructose solution induces hyperinsulinemia and hyperglycemia in hamsters. *Clinics* **2007**, *62*, 327–334. [CrossRef]
11. Bartkiene, E.; Sakiene, V.; Bartkevics, V.; Wiacek, C.; Rusko, J.; Lele, V.; Ruzauskas, M.; Juodeikiene, G.; Klupsaite, D.; Bernatoniene, J.; et al. Nutraceuticals in gummy candies form prepared from lacto-fermented lupine protein concentrates, as high-quality protein source, incorporated with *Citrus paradise* L. essential oil and xylitol. *Int. J. Food Sci. Technol.* **2018**, *53*, 2015–2025. [CrossRef]
12. Lê, K.A.; Ith, M.; Kreis, R.; Faeh, D.; Bortolotti, M.; Tran, C.; Tappy, L. Fructose overconsumption causes dyslipidemia and ectopic lipid deposition in healthy subjects with and without a family history of type 2 diabetes. *Am. J. Clin. Nutr.* **2009**, *89*, 1760–1765. [CrossRef] [PubMed]
13. Burton, P.; Lightowler, H.J. Influence of bread volume on glycaemic response and satiety. *Brit. J. Nutr.* **2006**, *96*, 877–882. [CrossRef] [PubMed]
14. Bae, I.Y.; Jun, Y.; Lee, S.; Lee, H.G. Characterization of apple dietary fibers influencing the in vitro starch digestibility of wheat flour gel. *LWT Food Sci. Technol.* **2016**, *65*, 158–163. [CrossRef]
15. Baeva, M.R.; Panchev, I.N.; Terzieva, V.V. Comparative study of texture of normal and energy reduced sponge cakes. *Nahrung* **2000**, *44*, 242–246. [CrossRef]
16. Karp, S.; Wyrwisz, J.; Kurek, M.A.; Wierzbicka, A. Combined use of cocoa dietary fibre and steviol glycosides in low-calorie muffins production. *Int. J. Food Sci. Technol.* **2017**, *52*, 944–953. [CrossRef]
17. Kulthe, A.A.; Pawar, V.D.; Kotecha, P.M.; Chavan, U.D.; Bansode, V.V. Development of high protein and low calorie cookies. *Int. J. Food Sci. Technol.* **2014**, *51*, 153–157. [CrossRef]
18. Livesey, G. Health potential of polyols as sugar replacers, with emphasis on low glycaemic properties. *Nutr. Res. Rev.* **2003**, *16*, 163–191. [CrossRef]
19. Oehme, A.; Wüst, M.; Wölwer-Rieck, U. Steviol glycosides are not altered during commercial extraction and purification processes. *Int. J. Food Sci. Technol.* **2017**, *52*, 2156–2162. [CrossRef]
20. Azevedo, B.M.; Morais-Ferreira, J.M.; Luccas, V.; Bolini, H.M. Bittersweet chocolates containing prebiotic and sweetened with stevia (*Stevia rebaudiana* Bertoni) with different Rebaudioside A contents: Multiple time-intensity analysis and physicochemical characteristics. *Int. J. Food Sci. Technol.* **2017**, *52*, 1731–1738. [CrossRef]
21. Wardy, W.; Jack, A.R.; Chonpracha, P.; Alonso, J.R.; King, J.M.; Prinyawiwatkul, W. Gluten-free muffins: Effects of sugar reduction and health benefit information on consumer liking, emotion, and purchase intent. *Int. J. Food Sci. Technol.* **2018**, *53*, 262–269. [CrossRef]
22. Gregersen, S.; Jeppesen, P.B.; Holst, J.J.; Hermansen, K. Antihyperglycemic effects of stevioside in type 2 diabetic subjects. *Metabolism* **2004**, *53*, 73–76. [CrossRef] [PubMed]
23. Goyal, S.K.; Goyal, R.K. Stevia (*Stevia rebaudiana*) a bio-sweetener: A review. *Int. J. Food Sci. Nutr.* **2010**, *61*, 1–10. [CrossRef] [PubMed]

24. Phimolsiripol, Y.; Siripatrawan, U.; Teekachunhatean, S.; Wangtueai, S.; Seesuriyachan, P.; Surawang, S.; Laokuldilok, T.; Regenstein, J.M.; Henry, C.J. Technological properties, *in vitro* starch digestibility and *in vivo* glycaemic index of bread containing crude malva nut gum. *Int. J. Food Sci. Technol.* **2017**, *52*, 1035–1041.
25. Quiles, A.; Llorca, E.; Schmidt, C.; Reißner, A.-M.; Struck, S.; Rohm, H.; Hernando, I. Use of berry pomace to replace flour, fat or sugar in cakes. *Int. J. Food Sci. Technol.* **2018**, *53*, 1579–1587. [CrossRef]
26. Manisha, G.; Soumya, C.; Indrani, D. Studies on interaction between stevioside, liquid sorbitol, hydrocolloids and emulsifiers for replacement of sugar in cakes. *Food Hydrocoll.* **2012**, *29*, 363–373. [CrossRef]
27. Chang, J.C.; Wu, M.C.; Liu, I.M.; Cheng, J.T. Increase of insulin sensitivity by stevioside in fructose-rich chow-fed rats. *Horm. Metab. Res.* **2005**, *37*, 610–616. [CrossRef]
28. O'Donnell, K.; Kearsley, M. *Sweeteners and Sugar Alternatives in Food Technology*, 2nd ed.; Wiley-Blackwell: Oxford, UK, 2012; pp. 215–218.
29. Wheeler, A.; Boileau, A.C.; Winkler, P.C.; Compton, J.C.; Prakash, I.; Jiang, X.; Mandarino, D.A. Pharmacokinetics of rebaudioside A and stevioside after single oral doses in healthy men. *Food Chem. Toxicol.* **2008**, *46*, S54–S60. [CrossRef]
30. Gao, J.; Brennan, M.A.; Mason, S.L.; Brennan, C.S. Effect of sugar replacement with stevianna and inulin on the texture and predictive glycaemic response of muffins. *Int. J. Food Sci. Technol.* **2016**, *51*, 1979–1987. [CrossRef]
31. Alencar, N.M.M.; de Morais, E.C.; Steel, C.J.; Bolini, H.M.A. Sensory characterisation of gluten-free bread with addition of quinoa, amaranth flour and sweeteners as an alternative for coeliac patients. *Int. J. Food Sci. Technol.* **2017**, *52*, 872–879. [CrossRef]
32. Abdel-Salam, A.M.; Ammar, A.S.; Galal, W.K. Evaluation and properties of formulated low calories functional yoghurt cake. *J. Food Agric. Environ.* **2009**, *7*, 218–221.
33. Edelstein, S.; Smith, K.; Worthington, A.; Gillis, N.; Bruen, D.; Kang, S.H.; Guiducci, G. Comparisons of Six New Artificial Sweetener Gradation Ratios with Sucrose in Conventional-Method Cupcakes Resulting in Best Percentage Substitution Ratios. *J. Culin. Sci. Technol.* **2007**, *5*, 61–74. [CrossRef]
34. Struck, S.; Jaros, D.; Brennan, C.S.; Rohm, H. Sugar replacement in sweetened bakery goods. *Int. J. Food Sci. Technol.* **2014**, *49*, 1963–1976. [CrossRef]
35. Röper, H.; Goossens, J. Erythritol, a new raw material for food and non-food applications. *Starch* **1993**, *45*, 400–405. [CrossRef]
36. Lin, S.D.; Hwang, C.F.; Yeh, C.H. Physical and Sensory Characteristics of Chiffon Cake Prepared with Erythritol as Replacement for Sucrose. *J. Food Sci.* **2003**, *68*, 2107–2110. [CrossRef]
37. Bornet, F.R. Undigestible sugars in food products. *Am. J. Clin. Nutr.* **1994**, *59*, 763S–769S. [CrossRef]
38. Lin, S.D.; Lee, C.C.; Mau, J.L.; Lin, L.Y.; Chiou, S.Y. Effect of Erythritol on Quality Characteristics of Reduced-Calorie Danish Cookies. *J. Food Qual.* **2010**, *33*, 14–26. [CrossRef]
39. Bornet, F.R.J.; Blayo, A.; Dauchy, F.; Slama, G. Plasma and Urine Kinetics of Erythritol after Oral Ingestion by Healthy Humans. *Regul. Toxicol. Pharm.* **1996**, *24*, S280–S285. [CrossRef]
40. Wang, Y.; Chen, L.; Li, Y.; Li, Y.; Yan, M.; Chen, K.; Xu, L. Efficient enzymatic production of rebaudioside A from stevioside. *Biosci. Biotech. Bioch.* **2016**, *80*, 67–73. [CrossRef]
41. Zahn, S.; Forker, A.; Krügel, L.; Rohm, H. Combined use of rebaudioside A and fibres for partial sucrose replacement in muffins. *LWT Food Sci. Technol.* **2013**, *50*, 695–701. [CrossRef]
42. AACC International. *Approved Methods of Analysis*, 9th ed.; AACC International: St. Paul, MN, USA, 1995; p. 76.
43. Martínez-Cervera, S.; Hera, E.; de la Sanz, T.; Gómez, M.; Salvador, A. Effect of using Erythritol as a Sucrose Replacer in Making Spanish Muffins Incorporating Xanthan Gum. *Food Bioprocess Tech.* **2012**, *5*, 3203–3216. [CrossRef]
44. Martínez-Cervera, S.; Sanz, T.; Salvador, A.; Fiszman, S.M. Rheological, textural and sensorial properties of low-sucrose muffins reformulated with sucralose/polydextrose. *LWT Food Sci. Technol.* **2012**, *45*, 213–220. [CrossRef]
45. Ghosh, S.; Sudha, M.L. A review on polyols: New frontiers for health-based bakery products. *Int. J. Food Sci. Nutr.* **2012**, *63*, 372–379. [CrossRef] [PubMed]
46. Morris, C.; Morris, G.A. The effect of inulin and fructo-oligosaccharide supplementation on the textural, rheological and sensory properties of bread and their role in weight management: A review. *Food Chem.* **2012**, *133*, 237–248. [CrossRef]

47. Rößle, C.; Ktenioudaki, A.; Gallagher, E. Inulin and oligofructose as fat and sugar substitutes in quick breads (scones): A mixture design approach. *Eur. Food Res. Technol.* **2011**, *233*, 167–181. [CrossRef]
48. Psimouli, V.; Oreopoulou, V. The Effect of Fat Replacers on Batter and Cake Properties. *J. Food Sci.* **2013**, *78*, C1495–C1502.
49. Ronda, F.; Gómez, M.; Blanco, C.A.; Caballero, P.A. Effects of polyols and nondigestible oligosaccharides on the quality of sugar-free sponge cakes. *Food Chem.* **2005**, *90*, 549–555. [CrossRef]
50. Alizadeh, M.; Azizi-Lalabadi, M.; Kheirouri, S. Impact of Using Stevia on Physicochemical, Sensory, Rheology and Glycemic Index of Soft Ice Cream. *FNS* **2014**, *5*, 390. [CrossRef]
51. Ishikawa, M.; Miyashita, M.; Kawashima, Y.; Nakamura, T.; Saitou, N.; Modderman, J. Effects of Oral Administration of Erythritol on Patients with Diabetes. *Regul. Toxicol. Pharm.* **1996**, *24*, S303–S308. [CrossRef]
52. Granfeldt, Y.; Eliasson, A.-C.; Björck, I. An Examination of the Possibility of Lowering the RGlycemic Index of Oat and Barley Flakes by Minimal Processing. *J. Nutr.* **2000**, *130*, 2207–2214. [CrossRef]
53. Anton, S.D.; Martin, C.K.; Han, H.; Coulon, S.; Cefalu, W.T.; Geiselman, P.; Williamson, D.A. Effects of stevia, aspartame, and sucrose on food intake, satiety, and postprandial glucose and insulin levels. *Appetite* **2010**, *55*, 37–43. [CrossRef] [PubMed]
54. Roberts, A.; Renwick, A.G. Comparative toxicokinetics and metabolism of rebaudioside A, stevioside, and steviol in rats. *Food Chem. Toxicol.* **2008**, *46*, S31–S39. [CrossRef] [PubMed]

© 2019 by the authors. Licensee MDPI, Basel, Switzerland. This article is an open access article distributed under the terms and conditions of the Creative Commons Attribution (CC BY) license (http://creativecommons.org/licenses/by/4.0/).

foods

MDPI

Article

Comprehensive Nutrition Review of Grain-Based Muesli Bars in Australia: An Audit of Supermarket Products

Felicity Curtain [1,*] and Sara Grafenauer [1,2]

1 Grains & Legumes Nutrition Council, Mount Street, North Sydney 2060, Australia
2 School of Medicine, University of Wollongong, Northfields Avenue, Wollongong 2522, Australia
* Correspondence: f.curtain@glnc.org.au; Tel.: +61-9394-8663

Received: 25 July 2019; Accepted: 23 August 2019; Published: 28 August 2019

Abstract: Muesli bars are consumed by 16% of children, and 7.5% of adults, and are classified as discretionary in Australian Dietary Guidelines, containing "higher fat and added sugars" compared with core food choices. This study aimed to provide a nutritional overview of grain-based muesli bars, comparing data from 2019 with 2015. An audit of muesli bars, grain-based bars, and oat slices was undertaken in January 2019 (excluding fruit, nut, nutritional supplement, and breakfast bars) from the four major supermarkets in metropolitan Sydney. Mean and standard deviation was calculated for all nutrients on-pack, including whole grain per serve and per 100g. Health Star Rating (HSR) was calculated if not included on-pack. Of all bars (n = 165), 63% were ≤ 600 kJ (268–1958 kJ), 12% were low in saturated fat, 56% were a source of dietary fibre, and none were low in sugar. Two-thirds (66%) were whole grain (≥8 g/serve), with an average of 10 g/serve, 16% of the 48 g Daily Target Intake. HSR featured on 63% of bars (average 3.2), with an overall HSR of 2.7. Compared to 2015, mean sugars declined (26.6 g to 23.7 g/100 g; $p < 0.001$), and 31% more bars were whole grain (109 up from 60 bars). Although categorised as discretionary, there were significant nutrient differences across grain-based muesli bars. Clearer classification within policy initiatives, including HSR, may assist consumers in choosing products high in whole grain and fibre at the supermarket shelf.

Keywords: muesli bars; grains; whole grain; dietary fibre; snack foods; nutrition

---

1. Introduction

'Muesli bar' is a generic term that refers to baked or cold-formed cereal-based snack bars, and may contain other ingredients such as fruit, nuts, seeds, chocolate, yoghurt, and a variety of other fillings and/or toppings [1]. They are a popular food in Australia, with consumption per capita considered the third highest worldwide, behind Canada and the USA [2]. An estimated 7.5% of Australian adults ate muesli bars the day prior to the 2011–12 Australian Health Survey, with consumption more common in younger age groups (16% of 4–13 year olds, compared to 12.8% of 14–18 year olds, and less than 8% of those aged 19–50 years) [3]. Their popularity with children was noted in a 2005 paper reviewing the lunchbox content of Australian school children, which found an estimated 41.8% of lunchboxes included a muesli/fruit bar, though this also included non-grain-based bars, excluded from this research [4].

Data from the 2011–12 Australian Health Survey found muesli bars contributed overall less than 1% of total energy, protein, fats, sugars, and dietary fibre to Australians aged 2 years and older [5]. However for females aged between 2–18 years, these figures were slightly higher; 1.1% energy, 1.2% total sugars, and 1.5% dietary fibre, and for males 2–18 years; 1.2% energy, 1% saturated fat, 1.4% total sugars, and 1.6% dietary fibre [5]. There is a lack of consensus on what constitutes a 'snack food', with definitions ranging from foods consumed between main meals or at specific times of day, food-type, or participant-described. Based on 'time of day' consumption, bars can be considered a snack food, generally eaten between main meals, and snacking of this kind has been linked with concern around increased risk of obesity and related chronic disease [6], though importantly, these health outcomes are multifactorial, with food choice and energy balance key in determining whether snacking is a healthful or harmful food behaviour [7,8].

Between 1995 and 2012, the prevalence and frequency of children snacking (defined as a single eating occasion between main meals) rose in Australia, with more than double the number of children snacking four or more times per day in 2012 [9]. Subsequently, the contribution of snacks to total energy intake significantly increased, from 24–30.5%. Foods consumed as snacks were a mix of traditional 'snack' foods such as sweet biscuits, cakes, fresh fruit, and 'meal' foods, such as bread and milk. Fruit and vegetable juice was the top contributor to energy from snacks in 1995, but did not appear in 2012, with pome fruit moving up as the top contributor. Muesli bars did not feature in the top snacks in 1995, but were number seven in 2007, and number nine in 2012, where they contributed an estimated 12.5% of total energy to snacks [9]. In Australian adults, cakes, muffins, scones, breads, and dairy milk were the three greatest contributors to energy from snacks, with 22% of total energy derived from snacking occasions [10]. While no data has reviewed changes in snacking habits among Australian adults, steady increases from 1977–2006 amongst adults in the USA mirror Australian children's results, contributing more kilojoules, mainly from discretionary foods like desserts, sugar sweetened beverages, and salty snacks [11].

The popularity of muesli bars, and increasing levels of consumption [9] have attracted attention from public health groups, government, and the media, not least since they are considered a 'discretionary' food in the Australian Dietary Guidelines, where their consumption is discouraged based on having "higher fat and added sugars" [12]. Importantly, they are not depicted in the accompanying Australian Guide to Healthy eating, which visually represents core and discretionary foods. Instead, muesli bars are listed in the longer form supplementary text, and are therefore hidden from view, so it is unclear how well understood their classification as discretionary is among consumers. Similarly, the New Zealand Eating and Activity Guidelines present muesli bars as an example of a 'highly processed' food that may be refined and contain added saturated fat, sugar, and salt [13], and the United Kingdom's Eat Well Guide cautions that cereal bars may have high levels of added sugars [14].

In 2018, proposed sugar reformulation targets for muesli bars were developed by The Healthy Food Partnership, an initiative established by the Australian Government in 2015, which aims to improve public health nutrition through several policy areas, including food reformulation [1]. Their inclusion was noteworthy, as they did not comply with the initial criteria (contributing significantly (≥1%) to sodium, sugars, and/or saturated fat in the Australian population's intake), instead being included based on their high level of consumption among children [1]. The proposed targets call for a "10% reduction in sugar across defined products containing over 28 g sugar/100 g, and a reduction in sugar to 25 g/100 g for products between 25–28 g sugar/100 g by the end of 2022". It is important to recognise that many companies have their own nutrition policies and commitments, as outlined in a 2018 Australian report, which found 16 of the 19 food companies surveyed included nutrition in their corporate strategy and had a commitment to product reformulation, while 11 out of 19 had committed to implementing the voluntary Health Star Rating (HSR) system [15].

The HSR is an interpretive Front of Pack Labelling system, first introduced in Australia and New Zealand in 2014, as a joint initiative between Government, public health, industry, and consumer groups. The system uses an algorithm to assign a star rating between 0.5–5 stars, and is intended to aid consumers in making healthier choices within categories [16,17]. The HSR algorithm rates foods on a per 100 g basis, considering both 'negative nutrients' (kilojoules, saturated fat, total sugars, and sodium), and 'positive' elements (fruit, vegetables, nuts and legumes, as well as protein and dietary fibre in some cases), which is then converted to a star rating [18]. Muesli bars were a key category of consideration in the ongoing HSR 5-year review, which noted they had received negative media attention based on products scoring "inappropriately high scores", despite their categorisation as discretionary foods [19].

However, grain-based muesli bars may also be a potential source of positive ingredients and nutrients within the diet pattern, particularly considering whole grain and dietary fibre content, which are promoted within Australian Dietary Guidelines [12]. Widespread evidence supports whole grains and whole grain foods for their protective health benefits, including lower total and cause-specific mortality, type 2 diabetes [20–24], weight gain [25], and colorectal cancer [26]. Globally, low whole grain intake has been recognised as the second greatest dietary risk factor for mortality (behind sodium), and the greatest dietary risk factor for morbidity, responsible for more than 80 million Disability-Adjusted-Life-Years [27]. Irrespective of its well-documented health benefits, whole grain intake in Australia is low, with follow up data from the Australian Health Survey recording median intake for children at 16.5 g per day, and adults at 21.2 g/day—both less than half of the established Daily Target Intake (DTI) of 48 g per day for adults, and between 32 and 40 g per day for children [28–30]. Equally, a large body of evidence points to the benefits of dietary fibre and its role in reducing chronic disease risk, yet most Australians fall short, with more than half of children, and more than 70% of adults not meeting their respective targets [31].

Due to their popularity and increasing consumption in Australia, muesli bars are often criticised and met with confusion regarding their nutritional value, with a particular focus on sugar content. This study aimed to provide an overview of the nutritional status of grain-based muesli bars on shelf including muesli bars, grain-based bars, and oat slices in Australian supermarkets, and provide a comparison of 2019 with 2015 data.

2. Materials and Methods

An audit of grain-based muesli bars was conducted January 2019, in four major supermarkets in metropolitan Sydney (Aldi, Coles, IGA, and Woolworths). Collectively, these supermarket chains make up more than 80% of total Australian market share, and were chosen in preference to smaller, independent grocery stores in an attempt to reflect food choices that the majority of Australians are faced with during food shopping [32]. This recognised process has been outlined in previously published research [33] and the same process was utilised in the data collection from 2015. Smartphones were used to capture all information on food packaging, including ingredient lists, Nutrition Information Panels (NIP), health and nutrition claims, HSR, and any additional logos and endorsements. Outlined in Table 1 below, products accounted for in the audit included muesli bars, grain-based bars, (including fruit-filled bars and twists, and those made from wheat, puffed rice, or other grains), and oat slices. Products were further categorised to determine whether they were specifically marketed towards children, by the presence of cartoons, promotions, or sporting figures, as described in previous research [34,35]. Products excluded were fruit-based bars, fruit leather/straps, nutritional supplement bars (e.g. protein/'low-carb' bars), nut/seed based bars, and breakfast bars/biscuits (e.g. those designed as a meal replacement, indicated in the product name), in line with exclusions within the Healthy Food Partnership proposed reformulation targets [1]. A supplementary internet search was conducted through supermarket websites and identified manufacturer websites using key words such as "snack bars", "muesli bars", "grain-based bars", "oat slices", and "snack bars", to ensure all products were captured.

Data from photographs taken at both timeframes (2015 and 2019) were transcribed into a Microsoft® Excel® spreadsheet (Version 2013, Redmond, Washington, DC, USA) for analysis. Information for the data entry included the NIP per serve and per 100 g, ingredients, percentage of whole grains, nutrition and health related claims, including whole grain, protein, dietary fibre, saturated fat, sugars, and sodium. Eligibility for products to make nutrition content claims was also assessed, in line with Food Standards Australia New Zealand and GLNCs Code of Practice for Whole Grain Ingredient Content Claims (The Code) [30], as well as proportion of products meeting the Healthy Food Partnership proposed reformulation targets for sugar reduction. HSR was not collected in 2015 as this was not on pack at this time. Where HSR was not featured on packaging, it was calculated for all products using the HSR website calculator [36]. A second, independent reviewer checked data for any inconsistencies and errors, and results were compared with 2015 data that followed the same process, to assess changes.

Table 1. Classification of categories.

Category	Description
Muesli bar	Baked or cold-formed bars where oats made up ≥5% of the product OR were one of the first five ingredients listed on the Nutrition Information Panel (NIP)
Grain-based bar	Baked or cold-formed bars where grain ingredient (s) (excluding oats) made up ≥5% of the product OR grains (excluding oats) were one of the first five ingredients listed on the NIP
Oat slice	Soft-baked bars with the word 'slice' in the product name, where oats made up ≥5% of the product OR were one of the first five ingredients listed on the NIP

Statistics

All data were checked for normality using Shapiro–Wilk test (IBM SPSS®, version 25.0, IBM Corp., Chicago, IL, USA) and mean and standard deviation were presented. As expected, there were missing values for dietary fibre and whole grain as these are often not presented unless a claim is being made on-pack, therefore dietary fibre and whole grain were analysed separately.

One-way ANOVA with post hoc Tukey analysis (IBM SPSS®, version 25.0, IBM Corp., Chicago, IL, USA) was used to compare differences per serve and per 100 g between (1) muesli bars, (2) grain-based bars, (including fruit-filled bars and twists, and those made from wheat, puffed rice, or other grains), and (3) oat slices for all available nutrients reported on-pack, including where relevant, dietary fibre, whole grain (g and %) and HSR (per 100 g). Independent samples *t*-test (IBM SPSS®, version 25.0, IBM Corp., Chicago, IL, USA) was used to compare whole grain and refined grain bars, which was defined according to each product's eligibility for registration with The Code (≥8 g whole grain per manufacturer serve), a method that has been described in previously published research [37]. *T*-tests were also used to determine difference in HSR for all products /100 g, between whole grain and refined grain categories and for data per 100 g from 2015 compared with 2019.

3. Results

Data from 165 bars were collected, including 96 muesli bars, 46 grain-based bars, and 23 oat slices from 18 manufacturers where the top three (Nestle Ltd., Kellogg (Aust) Pty. Ltd. and Carman's Fine Foods Pty. Ltd.), hold more than 60% market share (Retail World, December 2018) and have national distribution. Of these, 28 bars (17%) were identified as being specifically marketed towards children; these were predominantly grain-based bars (71%), with the remaining 8% muesli bars. Overall, mean serve size varied substantially between categories, with grain-based bars the smallest (27 g), followed by 35 g for muesli bars, and 55 g for oat slices.

There was a significant difference in nutrients including whole grain across all categories per serve and per 100 g (Tables 2 and 3). Post hoc Tukey analysis (per serve) comparing muesli bars and grain-based bars revealed no significant differences in saturated fat ($p = 0.181$), carbohydrate ($p = 0.365$), sugars ($p = 0.274$), and sodium ($p = 0.869$). Grain-based bars and oat slices were significantly different across all nutrients and whole grain content. Conversely, muesli bars and oat slices were the closest in composition for dietary fibre and whole grain ($p = 0.273$ and $p = 0.238$ respectively) with grain-based bars significantly lower ($p < 0.001$). Almost all (95%) grain-based bars met the Australian Dietary Guidelines recommendations of 600 kJ or less as a 'serve' of discretionary food, as well as 61% of muesli bars, but only 8% of oat slices.

Table 2. Nutrients per serve (mean & SD): muesli bars, grain-based bars, and oat slices including whole grain.

Nutrient Criteria	Muesli Bars (n = 96)	Grain-Based Bars (n = 46)	Oat Slices (n = 23)	p-Value	Total Bars (n = 165)
Serve Size (g)	35 ± 7.5	27 ± 7.0	55 ± 29.9	<0.001	35 ± 15.7
Energy (kJ)	614.4 ± 155.6	428.4 ± 91.6	1007.7 ± 565.6	<0.001	617.4 ± 301.0
Protein (g)	3.1 ± 1.7	1.5 ± 0.6	4.1 ± 2.5	<0.001	2.8 ± 1.9
Fat (g)	5.6 ± 2.9	2.3 ± 1.4	11.2 ± 7.0	<0.001	5.4 ± 4.4
Saturated Fat (g)	1.7 ± 1.2	1.1 ± 0.7	7.1 ± 4.3	<0.001	2.3 ± 2.7
Carbohydrate (g)	19.7 ± 3.9	18.0 ± 5.0	29.6 ± 15.8	<0.001	20.6 ± 7.9
Sugars (g)	7.3 ± 2.5	8.2 ± 3.4	12.0 ± 5.3	<0.001	8.2 ± 3.6
Dietary Fibre (g)	3.2 ± 1.6	1.6 ± 1.5	3.8 ± 2.4	<0.001	2.8 ± 1.9
Sodium (mg)	39.7 ± 44.1	43.5 ± 19.9	99.0 ± 61.4	<0.001	49.0 ± 46.4
Whole Grain (g)	14.2 ± 4.8	1.0 ± 2.8	16.1 ± 8.2	<0.001	10.7 ± 7.9

One Way ANOVA 95% CI.

Table 3. Nutrients, whole grain, and HSR/100 g (mean & SD) in muesli bars, grain-based bars, and oat slices.

Nutrient Criteria	Muesli Bars (n = 96)	Grain-Based Bars (n = 46)	Oat Slices (n = 23)	p-Value
Energy (kJ)	1770.9 ± 180.3	1633.4 ± 188.2	1817.9 ± 112.7	<0.001
Protein (g)	8.6 ± 3.2	5.4 ± 1.9	7.3 ± 1.0	<0.001
Fat (g)	15.4 ± 5.1	9.1 ± 5.8	20.0 ± 4.4	<0.001
Saturated fat (g)	5.0 ± 3.3	4.3 ± 3.7	12.9 ± 3.3	<0.001
Carbohydrate (g)	56.7 ± 8.3	67.6 ± 7.4	53.9 ± 3.9	<0.001
Sugars (g)	20.9 ± 5.6	29.8 ± 6.4	23.1 ± 4.3	<0.001
Dietary Fibre (g)	9.4 ± 5.1	5.8 ± 6.1	6.6 ± 0.8	<0.001
Sodium (mg)	112.2 ± 121.0	166.7 ± 69.9	174.5 ± 46.7	0.002
HSR	3.1 ± 1.1	2.4 ± 1.0	1.8 ± 0.4	<0.001
% Whole Grain	40.7 ± 11.8	4.8 ± 13.4	29.7 ± 4.6	<0.001

One Way ANOVA 95% CI.

Comparing per 100 g, post hoc Tukey analysis revealed no difference in saturated fat ($p = 0.558$) between muesli bars and grain-based bars although all other nutrients and HSR were significantly different ($p < 0.001$). Similarly, all nutrients were significantly different between grain-based bars and oat slices except sodium ($p = 0.952$) and although muesli bars are most similar to oat slices in terms of dietary fibre and whole grain content as noted earlier, there were significant differences in fat ($p = 0.001$), saturated fat ($p < 0.001$), sodium ($p = 0.009$), and HSR ($p = 0.001$). Muesli bars were highest in dietary fibre, contributing an average of 9.4 g/100 g, the lowest in sodium (112.2 mg/100 g), and had a significantly higher HSR (3.0). They also contained the highest percentage of whole grain ingredients (40.7%) compared with grain-based bars and oat slices. The average HSR for all products was 2.7, but was higher for the 63% of products that displayed it on-pack (3.2 stars) compared to those that did not (1.8 stars).

The overall results for bars specifically targeted towards children were similar to the averages for grain-based bars, with an average of 1659 kJ ± 120 per 100 g, 6.1 ± 3.3g protein, 9.8 ± 3.5 g total fat, 4.3 ± 2.8 g saturated fat, 67 ± 7.9 g carbohydrate, 26 ±8.1 g sugars, 6.1 ± 4 g dietary fibre, and 161 ± 78.1 mg sodium. Children's bars contained 19 ± 23.8% whole grain ingredients (contributing an average of 4.6 g to the 32–40 g Daily Intake Target for the 4–13 year old age group), and had an average HSR of 2.7 ± 1.1 stars, in line with the mean for the total snack bar category.

The percentage of products meeting nutrition claim criteria are presented in Table 4. More than half of muesli bars and oat slices were eligible for a 'contains whole grain' claim (compared to only 4% of grain-based bars), and 17% of oat slices were considered very high in whole grain. Six products did not report their percentage of whole grain ingredients, required to determine claim eligibility, so these were assumed as ineligible. Similar results were obtained for fibre claim eligibility, with 56% of the total category at least a source of fibre, mostly represented by muesli bars (69%), and oat slices (61%). The greatest proportion of grain-based bars were low in saturated fat (30%), compared to only 5% of muesli bars, and no oat slices. While none of the investigated bars were considered low in sugar, 48% overall met the most stringent proposed sugar reformulation target for muesli bars, (<25 g/100 g), and an additional 13% met the lower level proposed target of between 25–28 g sugar/100 g, with 29% falling outside the criteria.

Table 4. Percentage of products meeting claim criteria and proposed reformulation targets *.

	Muesli Bars (n = 96)	Grain-Based Bars (n = 46)	Oat Slices (n = 23)	Total Snack Bars (n = 165)
Eligible for WG claim (≥8 g/manufacturer serve)	90	4	91	66
Contains WG (≥8 g/manufacturer serve)	58	4	65	43
High in WG (16–24 g/manufacturer serve)	41	0	4	16
Very High in WG (≥24 g/manufacturer serve)	4	0	17	6
Source of Fibre (≥2–<4 g/serve)	69	26	61	56
Good Source of Fibre (≥4–<7 g/serve)	5	2	4	4
Excellent Source of Fibre (≥7 g/serve)	5	2	22	7
Low in Saturated Fat (≤1.5 g/100 g)	5	30	0	12
Low in Sugar (≤5 g/100 g)	0	0	0	0
Meets Proposed Sugar Reformulation Target 25–28 g/100 g *	9	11	4	9
Meets Proposed Sugar Reformulation Target <25 g/100 g *	78	24	65	61

* Healthy Food Partnership proposed reformulation targets (September 2018).

As outlined in Table 5, bars categorised as whole grain (≥8 g per manufacturer serve) were significantly higher in energy, total fat, and dietary fibre, and lower in sugars and sodium than refined grain bars. Interestingly, there was no significant difference noted in HSR between whole grain and refined grain bars, with 0.7 star between those categorised as whole grain and the remaining 'non-whole grain bars' which were categorised as refined grain bars.

Table 5. Whole grain versus refined grain nutrients (per 100 g) (mean and SD).

NIP	Whole Grain * (n = 109)	Refined Grain ** (n = 56)	p-Value
Energy (kJ)	1772.6 ± 171.1	1673.8 ± 199.6	0.044
Protein (g)	8.4 ± 2.9	5.9 ± 2.3	0.384
Fat (g)	16.1 ± 5.2	10.9 ± 7.0	0.034
Saturated Fat (g)	6.2 ± 4.4	5.3 ± 4.3	0.389
Carbohydrate (g)	56.2 ± 7.4	65.4 ± 9.3	0.059
Sugars (g)	20.9 ± 5.1	29.1 ± 6.6	0.043
Dietary Fibre (g)	8.6 ± 4.2	6.6 ± 6.7	0.008
Sodium (mg)	119.5 ± 111.0	168.4 ± 82.2	0.005
HSR	3.1 ± 1.1	2.4 ± 1.0	0.075

Independent samples t-test 95% CI. * Based on eligibility for registration with GLNCs Code of Practice for Whole Grain Ingredient Content Claims (≥8 g per manufacturer serve). ** Includes six bars that did not report percentage of whole grain ingredients.

In regards to other on-pack claims, 'No artificial colours/flavours/preservatives' was the most common claim made on packaging, featuring on almost three-quarters (73%) of the total category, and on 91% of oat slices, 80% of grain-based bars, and 66% of muesli bars. More than half made a dietary fibre claim (56%), including 60% of both oat slices and muesli bars, and 30% of grain-based bars. Similarly, 49% made a whole grain claim on-pack, mainly seen on oat slices (70%), and muesli bars (68%), with only 9% of grain-based bars making this claim. An additional 28 products were eligible, but did not make a whole grain claim.

Compared with 2015 (Table 6), 3.5% fewer bars were captured (171 versus 165), with apparent growth in the number of muesli bars (82 to 96 products), and oat slices (18 to 23 products), but a decline in grain-based bars (71 to 46 products), these being the most nutritionally poor products within the category. Over time, there was a significant decrease in total sugars from 26.6 g/100 g to 23.7 g/100 g ($p < 0.001$) across the total category in the four years since 2015, largely attributed to muesli bars, containing 4.2 g/100 g less sugars, while grain-based bars remained stable, and oat slices decreased by 1.1 g/100 g. The proportion of whole grain bars within the category increased, from 35 to 66% in four years (60/171 up to 109/165 bars). HSR data was not captured in 2015 due to the system being newly introduced, so no comparison of this metric over time was possible.

Table 6. Comparison of nutrients and whole grain in total bars between 2015 and 2019 per 100g (mean and SD).

Nutrient Criteria	Total Bars 2015 (n = 171)	Total Bars 2019 (n = 165)	p-Value
Energy (kJ)	1700 ± 179.9	1739.1 ± 186.7	0.049
Protein (g)	6.6 ± 2.1	7.5 ± 3.0	0.001
Fat (g)	13.1 ± 6.2	14.1 ± 6.2	0.089
Saturated Fat (g)	5.7 ± 4.4	5.9 ± 4.4	0.610
Carbohydrate (g)	62.3 ± 8.4	59.3 ± 9.2	0.002
Sugars (g)	26.6 ± 7.2	23.7 ± 6.8	<0.001
Dietary Fibre (g)	6.6 ± 4.1	7.9 ± 5.2	0.203
Sodium (mg)	143.1 ± 104.5	136.1 ± 104.5	0.540
% Whole Grain	30.0 ± 15.0	38.8 ± 11.2	0.009

Independent samples t-test 95% CI.

4. Discussion

Despite their widespread popularity, consumption of grain-based muesli bars are discouraged by the Australian Dietary Guidelines based on their classification as a discretionary food. This study aimed to provide a comprehensive overview of the nutritional status of grain-based muesli bars on shelf in Australian supermarkets, compared to data collected in 2015.

Overall, wide nutrient ranges were demonstrated between and within the categories examined although muesli bars are treated as a homogenous category in food policy and in advice to consumers. A major factor influencing these differences was the range in average serve sizes, with oat slices more than double that of grain-based bars. Serve size discrepancy may be a point of confusion for shoppers, as nutrient content of the smaller sized grain-based bars may appear more favourable, yet these were the highest in some nutrients of concern on a per 100 g basis. Conversely, oat slices are larger and appear the highest in some positive nutrients per serve, but not when compared per 100 g. This may suggest that the nutrition features of bars may be difficult to compare using the per serve nutrition information at the supermarket shelf. This has been previously described as 'health framing', whereby the impression of a healthier product may lead to overconsumption, however as all bars examined were individually wrapped and therefore portion controlled, this may be less of a concern than in other snack food categories such as cakes and biscuits. These findings are consistent with prior research in Australia which found significant variability in manufacturer serve size within both discretionary [38,39], and core food groups [40,41], and are partly explained by the lack of regulation around standard serving sizes in Australia, which is determined by food manufacturers [40].

Differing ingredients were also a major factor influencing variations in nutrition profile and serve size. Many grain-based bars consisted of puffed or flaked grains (such as corn or rice), which were likely lighter in weight than whole grains, more commonly found in muesli bars and oat slices. Oat slices often contained butter and coconut, both known for their high levels of saturated fat. Additionally, muesli bars and oat slices were all based on oats, which are unique among grains for their higher fat content (6–8%, compared to 2–3% in other grains [42]). The difference in ingredients provides basis for considering further differentiation within this category and at the same time, questions the broad categorisation of 'muesli bars' within the discretionary food group.

Almost one in five bars (17%) in 2019 were specifically marketed towards children, and these were mainly within the grain-based bars category (which are smaller and often made with puffed grains). Generally, these were less nutritious options, being lower in protein, dietary fibre, and whole grain, and higher in sugar than the category on average. Previous research has echoed this finding, with the products designed to appeal to children generally higher in some negative nutrients [34]. Encouragingly, their nutritional value was reflected in the average HSR of less than 3 stars, which has been determined as a cut off point for consumers identifying a food as unhealthy [43].

'Snackification', or the demand for convenience foods to suit modern lifestyles may drive continued innovation and reformulation. New Nutrition Business identified snacking as a key driver of food choice in 2018 and 2019, pointing to examples of manufacturers reinventing foods that were once impossible to eat on-the-go, such as peanut butter in portioned sachets and microwave porridge in individual pots, possibly increasing market competition for muesli bars as traditional snack foods [44]. When considering the top three contributors to adults (19–70+ years) discretionary food intake, the Australian Institute of Health & Welfare's 2018 Nutrition Across the Life Stages report listed alcohol, cakes/muffins/pastries, and soft drinks [45]. Similarly, a 2017 review analysing Australian children's discretionary food intake identified cakes/muffins/slices (4.2%), sweet biscuits (2.9%), and potato crisps/similar snacks (2.7%) as the top contributors to total energy, and the greatest contributors to added sugar were sugar-sweetened soft drinks (18.6%), cakes, muffins, and slices (10.6%), and cordials (6.7%). Conversely, 'sweet snack bars' (which included muesli/cereal bars, and fruit/nut/seed bars) contributed only 1.2% to total energy, and 1.6% added sugars [46]. When this is considered in the context of a typical Australian school lunchbox, including "about one sandwich, two biscuits, a piece of fruit, a snack of either a muesli/fruit bar or some other packaged snack, and a drink of fruit juice/cordial or water" [4], the particular focus on muesli bars as a food of concern may need to be reassessed against the full range of options that could be included in this meal occasion. Discretionary foods such as biscuits, cakes, potato chips, and cordial offer minimal nutritional benefits, so encouraging healthier options within the muesli bar category, alongside core foods in preference to these may be more beneficial advice to consumers and parents who are already under pressure to provide convenient, nutritious snacks.

Comparisons with 2015 data (in Table 6) are suggestive of improvements in terms of added sugars and whole grain content made by food industry. Reformulation aims to improve the nutritional content of manufactured foods, either by increasing beneficial nutrients, or reducing risk-associated nutrients. Often, manufacturers make modest nutritional changes over a period of time to allow consumers' tastes to adjust accordingly, referred to as "health by stealth" [47], but in recent years Australian muesli bar manufacturers have openly shared efforts to reduce salt, fat, sugar, and increase dietary fibre [48]. There is evidence to show reduction targets are effective, with a 2018 review of voluntary sodium reduction targets in soup demonstrating a 6% reduction in sodium levels in soup products between 2011 and 2014, with 67–74% of products compliant with targets [49]. Similarly, Australia's National Heart Foundation has reported significant reductions in line with targets set by the Food and Health Dialogue, such as 10% less sodium in bread and processed meats, and 32% less sodium in breakfast cereals [50], indicating that proposed targets set by the Healthy Food Partnership may encourage further improvements in the added sugars content of muesli bars.

Authors of the 2017 Global Burden of Disease study speculated that dietary policies focused on promoting consumption of whole grains, fruits. and vegetables, and other core food groups may have a greater effect than policies targeting excess consumption of sugar and fat [27]. Within the current study, whole grain bars were clearly identified as a healthier option overall, providing more protective nutrients, and fewer negative nutrients than refined grain bars. Across categories, the majority of oat slices and muesli bars were whole grain (≥8 g per manufacturer serve), and provided the equivalent of at least 30% of an adult's 48 g Daily Target Intake for whole grain, and up to half of a child's daily whole grain requirement (32–40 g/day) [30]. In light of this, whole grain bars may present a convenient, portion controlled, and accepted vehicle for whole grain, and their consumption over refined grain bars could aid in bridging the significant gap in consumption. Unlike other nutrients, whole grain claims are not regulated by Food Standards Australia New Zealand, but are instead encouraged through

GLNCs voluntary Code of Practice for Whole Grain Ingredient Content Claims (The Code), introduced in Australia and New Zealand in 2013 to encourage evidence-based promotion of whole grain foods. GLNC utilises audits of grain-based foods to monitor the operation of The Code and provide feedback to industry as necessary. While 60% of eligible bars were registered with The Code, its voluntary nature, and the fact that the percentage of whole grain ingredients is not mandatory in the ingredients list means deciphering which are whole grain options is not always clear to consumers. This was highlighted by the six bars identified that contained whole grain ingredients (such as rolled oats, and whole grain wheat), but did not report their percentages, so it was unclear whether they met The Code's whole grain criteria. Encouragingly, the number of whole grain bars have increased by 31% since 2015, suggesting positive changes have been made by manufacturers to existing products, new whole grain products have been added to the market due to consumer demand, or that labelling has been updated to more clearly communicate whole grain content.

The variability in nutrients supplied within the grain-based muesli bar category, combined with their popularity, may point towards education as the more powerful tool in supporting consumers to choose healthier products, in preference to discouraging consumption. The concept of 'knowledge-is-power' has been explored in previous research, with a review from the USA determining consumers with greater nutrition knowledge were more likely to consult nutrition labels, which may lead to healthier food choices [51]. The HSR attempts to clarify complex nutrition information and arm consumers with the knowledge to make healthier choices within food categories, and has been shown to perform well in directing consumers towards healthier, higher-scoring foods [43,52,53]. HSR scores for the bars category ranged from 1–5 stars, yet there was no significant difference between refined and whole grain varieties, with only 0.7 of a star between products. This finding highlights a shortcoming of the algorithm used to assign products a star rating, and builds on previous research that demonstrated an inability to differentiate whole grain and refined grain breads, breakfast cereals, rice, and flour products, as it does not directly account for, or reward foods for whole grain content [37]. There is a clear opportunity to refine the HSR by recognising whole grain as a positive food component, which could play a role in discerning healthier food choices across numerous categories, including muesli bars. However, to meet its objective of simple nutrition comparisons within categories, widespread uptake of a voluntary front-of-pack labelling system such as the HSR is required. Almost two-thirds (63%) of bars examined displayed a HSR, comparatively higher than overall uptake, which is estimated at 28% [16]. Consistent with existing literature, bars displaying a HSR tended to have higher scores, suggesting the system may be used strategically within and across brands [16,54]. Conversely, industry appear to be using the HSR as an incentive to improve a product's nutritional value, with recent studies in Australia and New Zealand identifying upwards of 83% of products displaying a HSR had been reformulated to increase their score [54,55].

Strengths of this study include its comprehensive nature, and to our knowledge, it is the first study that has reviewed muesli bars on shelf in Australia, with a comparison made to previously collected data. Also, where HSR was not provided, we calculated this for a more accurate representation of HSR across the category. However, there were some limitations. The research was focused only on grain-based bars, excluding others—such as nut bars and protein and low-carb bars—which may also be consumed as snacks though to a lesser extent than muesli bars [56]. While all efforts were made to capture the category in its entirety, differences may exist between geographic areas. As previously stated, reporting of dietary fibre and whole grain within the ingredients and Nutrition Information Panel is not mandatory in the absence of an on-pack claim, so was not always declared, and thus there was some missing data. Finally, we did not conduct an independent nutrition analysis, and were reliant on manufacturer information.

5. Conclusions

Although categorised as discretionary, there are significant nutrient differences across grain-based muesli bars, with well-chosen bars providing valuable amounts of whole grain and dietary fibre.

Muesli bars are a widely consumed snack food, particularly among younger age groups in Australia, yet their contribution and role in the diet is controversial, based on their classification at discretionary by Australian Dietary Guidelines. This study demonstrated significant variation between and within the category, with the whole grain options emerging as more nutritious compared to refined grain bars, and an indication of sugar reduction since 2015. Within a balanced diet, it is clear that some muesli bars can offer a convenient and nutritious snack, with many bars providing around 30% of an adult's, and up to half of a child's daily requirement for whole grain, and more than half of all products are at least a source of fibre. Both whole grains and dietary fibre are encouraged within Dietary Guidelines yet intakes across age groups tend to fall short of dietary targets. The current HSR algorithm does not appear to be overly favouring muesli bars (with an overall score of 2.7), and instead, could be improved to capture and differentiate whole grain options. Ongoing promotion of the higher HSR scoring bars, alongside proposed voluntary sugar reformulation targets and trends such as snackification, may be suggestive of opportunities and incentives for manufacturers to further improve the current range of products. Clearer classification within policy initiatives utilising evidence-based assessment of available products may help refine advice from healthcare professionals, and may be key in providing better direction for consumers to make healthier and acceptable snack food and lunchbox choices.

Author Contributions: Conceptualization, F.C. and S.G.; Methodology, F.C.; Formal analysis, S.G.; Original draft preparation, F.C.; Review and editing, S.G.

Funding: This research received no external funding but was supported by the Grains & Legumes Nutrition Council, a not-for-profit charity.

Acknowledgments: Thanks to James Sze, Student Dietitian from the University of Wollongong, NSW, who was involved in data collection as part of his university studies, and to Joanna Russell for statistical advice.

Conflicts of Interest: S.G. and F.C. are employed by the Grains & Legumes Nutrition Council, a not-for-profit charity.

References

1. Department of Health; Australian Government. Healthy Food Partnership Voluntary Food Reformulation Targets-Public Consultation. Available online: https://consultations.health.gov.au/population-health-and-sport-division-1/hfp-reformulation/ (accessed on 10 July 2019).
2. Mintel. A Year of Innovation in Snack Bars. 2019. Available online: https://clients.mintel.com/report/a-year-of-innovation-in-snack-bars-2019 (accessed on 10 July 2019).
3. Australian Bureau of Statistics. 4364.0.55.007-Australian Health Survey: Nutrition First Results-Foods and Nutrients, 2011–2012. Available online: http://www.abs.gov.au/ausstats/abs@.nsf/lookup/4364.0.55.007main+features12011-12 (accessed on 9 July 2019).
4. Sanigorski, A.M.; Bell, A.C.; Kremer, P.J.; Swinburn, B.A. Lunchbox contents of Australian school children: Room for improvement. *Eur. J. Clin. Nutr.* **2005**, *59*, 1310–1316. [CrossRef] [PubMed]
5. Australian Bureau of Statistics. 4364.0.55.012-Australian Health Survey: Consumption of Food Groups from the Australian Dietary Guidelines, 2011–2012. Available online: http://www.abs.gov.au/ausstats/abs@.nsf/Lookup/by%20Subject/4364.0.55.007~{}2011-12~{}Main%20Features~{}Cereals%20and%20cereal%20products~{}720 (accessed on 10 July 2019).
6. Wang, D.; van der Horst, K.; Jacquier, E.F.; Afeiche, M.C.; Eldridge, A.L. Snacking Patterns in Children: A Comparison between Australia, China, Mexico, and the US. *Nutrients* **2018**, *10*, 198. [CrossRef]
7. Mattes, R. Snacking: A cause for concern. *Physiol. Behav.* **2018**, *193*, 279–283. [CrossRef] [PubMed]
8. Mielmann, A.; Brunner, T.A. Consumers' snack choices: Current factors contributing to obesity. *Br. Food J.* **2019**, *121*, 347–358. [CrossRef]
9. Fayet-Moore, F.; Peters, V.; McConnell, A.; Petocz, P.; Eldridge, A.L. Weekday snacking prevalence, frequency, and energy contribution have increased while foods consumed during snacking have shifted among Australian children and adolescents: 1995, 2007 and 2011–12 National Nutrition Surveys. *Nutr. J.* **2017**, *16*, 65. [CrossRef] [PubMed]
10. Fayet-Moore, F.F.; McConnell, A.A.; Keighley, T.T. Adult snacking in Australia: Understanding who, what, when, and how much. *J. Nutr. Intermed. Metab.* **2016**, *4*, 15. [CrossRef]

11. Piernasm, C.; Popkin, B.M. Snacking Increased among U.S. Adults between 1977 and 2006. *J. Nutr.* **2009**. [CrossRef]
12. NHMRC. Australian Dietary Guidelines. Available online: https://www.nhmrc.gov.au/_files_nhmrc/file/publications/n55_australian_dietary_guidelines1.pdf (accessed on 9 July 2019).
13. New Zealand Ministry of Health. *Eating and Activity Guidelines for New Zealand Adults*; The Ministry of Health: Wellington, New Zealand, 2015.
14. England, P.H. The Eatwell Guide. Available online: https://assets.publishing.service.gov.uk/government/uploads/system/uploads/attachment_data/file/742750/Eatwell_Guide_booklet_2018v4.pdf (accessed on 10 July 2019).
15. Sacks, G.; Robinson, E. *For Informas. Inside Our Food and Beverage Manufacturers: Assessment of Company Policies and Commitments Related to Obesity Prevention and Nutrition, Australia*; Deakin University: Melbourne, Australia, 2018.
16. Jones, A.; Shahid, M.; Neal, B. Uptake of Australia's Health Star Rating System. *Nutrients* **2018**, *10*, 997. [CrossRef]
17. Commonwealth of Australia. About Health Star Ratings. Available online: http://healthstarrating.gov.au/internet/healthstarrating/publishing.nsf/Content/About-health-stars (accessed on 12 July 2019).
18. Jones, A.; Rådholm, K.; Neal, B. Defining 'Unhealthy': A Systematic Analysis of Alignment between the Australian Dietary Guidelines and the Health Star Rating System. *Nutrients* **2018**, *10*, 501. [CrossRef]
19. Mpconsulting. Five Year Review of the Health Star Rating System—Consultation Paper: Options for System Enhancement. Available online: http://www.healthstarrating.gov.au/internet/healthstarrating/publishing.nsf/Content/news-20181510/$File/HSR%20System%20Consultation%20Paper%20-%20October%202018.pdf (accessed on 10 July 2019).
20. Wu, H.; Flint, A.J.; Qi, Q.; van Dam, R.M.; Sampson, L.A.; Rimm, E.B.; Holmes, M.D.; Willett, W.C.; Hu, F.B.; Sun, Q. Association Between Dietary Whole Grain Intake and Risk of Mortality: Two Large Prospective Studies in US Men and Women. *JAMA Intern. Med.* **2015**. [CrossRef]
21. Aune, D.; Keum, N.; Giovannucci, E.; Fadnes, L.T.; Boffetta, P.; Greenwood, D.C.; Tonstad, S.; Vatten, L.J.; Riboli, E.; Norat, T. Whole grain consumption and risk of cardiovascular disease, cancer, and all cause and cause specific mortality: Systematic review and dose-response meta-analysis of prospective studies. *Br. Med. J.* **2016**, *353*, 2716. [CrossRef] [PubMed]
22. Zhang, B.; Zhao, Q.; Guo, W.; Bao, W.; Wang, X. Association of whole grain intake with all-cause, cardiovascular, and cancer mortality: A systematic review and dose-response meta-analysis from prospective cohort studies. *Eur. J. Clin. Nutr.* **2018**, *72*, 57–65. [CrossRef] [PubMed]
23. Zong, G.; Gao, A.; Hu, F.B.; Sun, Q. Whole Grain Intake and Mortality From All Causes, Cardiovascular Disease, and Cancer: A Meta-Analysis of Prospective Cohort Studies. *Circulation* **2016**, *133*, 2370–2380. [CrossRef] [PubMed]
24. Li, B.; Zhang, G.; Tan, M.; Zhao, L.; Jin, L.; Tang, X.; Jiang, G.; Zhong, K. Consumption of whole grains in relation to mortality from all causes, cardiovascular disease, and diabetes: Dose-response meta-analysis of prospective cohort studies. *Medicine (Baltim.)* **2016**, *95*, 4229. [CrossRef] [PubMed]
25. Maki, K.C.; Palacios, O.M.; Koeche, K.; Sawicki, C.M.; Livingston, K.A.; Bell, M.; Cortes, H.M.; McKeown, N.M. The Relationship between Whole Grain Intake and Body Weight: Results of Meta-Analyses of Observational Studies and Randomized Controlled Trials. *Nutrients* **2019**, *11*, 1245. [CrossRef] [PubMed]
26. Aune, D.; Chan, D.S.; Lau, R.; Vieira, R.; Greenwood, D.C.; Kampman, E.; Norat, T. Dietary fibre, whole grains, and risk of colorectal cancer: Systematic review and dose-response meta-analysis of prospective studies. *BMJ* **2011**, *343*, 6617. [CrossRef] [PubMed]
27. GBD 2017 Diet Collaborators. Health effects of dietary risks in 195 countries, 1990–2017: A systematic analysis for the Global Burden of Disease Study 2017. *Lancet* **2019**, *393*, 1958–1972. [CrossRef]
28. Griffiths, T. Towards an Australian 'daily target intake' for wholegrains. *Food Aust.* **2007**, *59*, 600–601.
29. Galea, L.; Beck, E.; Probst, Y.; Cashman, C. Whole grain intake of Australians estimated from a cross-sectional analysis of dietary intake data from the 2011-13 Australian Health Survey. *Public Health Nutr.* **2017**, *20*, 2166–2172. [CrossRef]
30. GLNC. Code of Practice for Whole Grain Ingredient Content Claims. Available online: http://www.glnc.org.au/codeofpractice/ (accessed on 25 July 2019).

31. Fayet Moore, F.; Cassettari, T.; Tuck, K.; McConnell, A.; Petocz, P. Dietary Fibre Intake in Australia. Paper I: Associations with Demographic, Socio-Economic, and Anthropometric Factors. *Nutrients* **2018**, *10*, 599. [CrossRef]
32. Roy Morgan. Woolworths and Aldi Grow Grocery Market Share in 2018. Available online: http://www.roymorgan.com/findings/7936-australian-grocery-market-december-2018-201904050426 (accessed on 8 August 2019).
33. Grafenauer, S.; Curtain, F. An Audit of Australian Bread with a Focus on Loaf Breads and Whole Grain. *Nutrients* **2018**, *10*, 1106. [CrossRef]
34. Mehta, K.; Phillips, C.; Ward, P.; Coveney, J.; Handsley, E.; Carter, P. Marketing foods to children through product packaging: Prolific, unhealthy and misleading. *Public Health Nutr.* **2012**, *15*, 1763–1770. [CrossRef] [PubMed]
35. Chapman, K.; Nicholas, P.; Banovic, D.; Supamaniam, R. The extent and nature of food promotion directed to children in Australian supermarkets. *Health Promot. Int.* **2006**, *21*. [CrossRef] [PubMed]
36. Commonwealth of Australia. Health Star Rating Calculator. Available online: http://www.healthstarrating.gov.au/internet/healthstarrating/publishing.nsf/Content/online-calculator#/step/1 (accessed on 1 July 2019).
37. Curtain, F.; Grafenauer, S.J. Health Star Rating in Grain Foods—Does It Adequately Differentiate Refined and Whole Grain Foods? *Nutrients* **2019**, *11*, 415. [CrossRef] [PubMed]
38. Haskelberg, H.; Neal, B.; Dunford, E.; Flood, V.; Rangan, A.; Thomas, B.; Cleanthous, X.; Trevena, H.; Zheng, J.M.; Louie, J.C.Y.; et al. High variation in manufacturer-declared serving size of packaged discretionary foods in Australia. *Br. J. Nutr.* **2016**, *115*, 1810–1818. [CrossRef] [PubMed]
39. Watson, W.L.; Kury, A.; Wellard, L.; Hughes, C.; Dunford, E.; Chapman, K. Variations in serving sizes of Australian snack foods and confectionery. *Appetite* **2016**, *96*, 32–37. [CrossRef] [PubMed]
40. Cleanthous, X.; Mackintosh, A.M.; Anderson, S. Comparison of reported nutrients and serve size between private label products and branded products in Australian supermarkets. *Nutr. Diet.* **2011**, *68*, 120–126. [CrossRef]
41. Yang, S.; Gemming, L.; Rangan, A. Large Variations in Declared Serving Sizes of Packaged Foods in Australia: A Need for Serving Size Standardisation? *Nutrients* **2018**, *10*, 139. [CrossRef]
42. Decker, E.A.; Rose, D.J.; Stewart, D. Processing of oats and the impact of processing operations on nutrition and health benefits. *Br. J. Nutr.* **2014**, *112*, S58–S64. [CrossRef] [PubMed]
43. Talati, Z.; Pettigrew, S.; Kelly, B.; Ball, K.; Dixon, H.; Shilton, T. Consumers' responses to front-of-pack labels that vary by interpretive content. *Appetite* **2016**, *101*, 205–213. [CrossRef]
44. Mellentin, J. *10 Key Trends in Food, Nutrition & Health 2019*; New Nutrition Business: London, UK, 2018.
45. Australian Institute of Health and Welfare. *Nutrition Across the Life Stages*; Australian Institue of Health and Weldare: Canberra, Australia, 2018.
46. Johnson, B.J.; Bell, L.K.; Zarnowiecki, D.; Rangan, A.M.; Golley, R.K. Contribution of Discretionary Foods and Drinks to Australian Children's Intake of Energy, Saturated Fat, Added Sugars and Salt. *Children* **2017**, *4*, 104. [CrossRef] [PubMed]
47. Regan, A.; Potvin Kent, M.; Raats, M.M.; McConnon, A.; Wall, P.; Dubois, L. Applying a Consumer Behavior Lens to Salt Reduction Initiatives. *Nutrients* **2017**, *9*, 901. [CrossRef] [PubMed]
48. Han, E. Uncle Tobys Cuts Fat, Salt and Sugar from Muesli Bars to Boost Health Star Rating. Available online: https://www.smh.com.au/business/companies/uncle-tobys-cuts-fat-salt-and-sugar-from-museli-bars-to-boost-health-star-rating-20151013-gk7n4j.html (accessed on 15 July 2019).
49. Levi, R.; Probst, Y.; Crino, M.; Dunford, E. Evaluation of Australian soup manufacturer compliance with national sodium reduction targets. *Nutr. Diet.* **2018**, *75*, 200–205. [CrossRef] [PubMed]
50. National Heart Foundation of Australia. *Report on the Evaluation of the Nine Food Categories for Which Reformulation Targets Were Set under the Food and Health Dialogue*; National Heart Foundation: Victoria, Australia, 2016.
51. Soederberg Miller, L.M.; Cassady, D.L. The effects of nutrition knowledge on food label use. A review of the literature. *Appetite* **2015**, *92*, 207–216. [CrossRef] [PubMed]
52. Actona, R.; Vanderleeb, L.; Hammond, D. Influence of front-of-package nutrition labels on beverage healthiness perceptions: Results from a randomized experiment. *Prev. Med.* **2018**, *115*, 83–89. [CrossRef] [PubMed]

53. Neal, B.; Crino, M.; Dunford, E.; Gao, A.; Greenland, R.; Li, N.; Ngai, J.; Ni Mhurchu, C.; Pettigrew, S.; Sacks, G.; et al. Effects of Different Types of Front-of-Pack Labelling Information on the Healthiness of Food Purchases—A Randomised Controlled Trial. *Nutrients* **2017**, *9*, 1284. [CrossRef] [PubMed]
54. Morrison, H.; Meloncelli, N.; Pelly, F.E. Nutritional quality and reformulation of a selection of children's packaged foods available in Australian supermarkets: Has the Health Star Rating had an impact? *Nutr. Diet.* **2019**, *76*, 296–304. [CrossRef] [PubMed]
55. Ni Mhurchu, C.; Eyles, H.; Choi, Y.-H. Effects of a Voluntary Front-of-Pack Nutrition Labelling System on Packaged Food Reformulation: The Health Star Rating System in New Zealand. *Nutrients* **2017**, *9*, 918. [CrossRef]
56. Mpconsulting. Five Year Review of the Health Star Rating (HSR) System: Snack Bar. Available online: http://www.mpconsulting.com.au/wp-content/uploads/2018/10/Snack-bars.pdf (accessed on 22 July 2019).

© 2019 by the authors. Licensee MDPI, Basel, Switzerland. This article is an open access article distributed under the terms and conditions of the Creative Commons Attribution (CC BY) license (http://creativecommons.org/licenses/by/4.0/).

foods

MDPI

Article

Gluten-Free Alternative Grains: Nutritional Evaluation and Bioactive Compounds

Serena Niro [1], Annacristina D'Agostino [1], Alessandra Fratianni [1,*], Luciano Cinquanta [2] and Gianfranco Panfili [1]

1 Dipartimento di Agricoltura, Ambiente e Alimenti, Università degli Studi del Molise, Via De Sanctis, 86100 Campobasso, Italy; serena.niro@unimol.it (S.N.); a.dagostino@studenti.unimol.it (A.D.A.); panfili@unimol.it (G.P.)

2 Dipartimento Scienze Agrarie, Alimentari e Forestali, Università di Palermo, Viale delle Scienze 4, 90128 Palermo, Italy; luciano.cinquanta@unipa.it

* Correspondence: fratianni@unimol.it

Received: 7 May 2019; Accepted: 10 June 2019; Published: 12 June 2019

Abstract: Interest in gluten-free grains is increasing, together with major incidences of celiac disease in the last years. Since to date, knowledge of the nutritional and bioactive compounds profile of alternative gluten-free grains is limited, we evaluated the content of water-soluble (thiamine and riboflavin) and liposoluble vitamins, such as carotenoids and tocols (tocopherols and tocotrienols), of gluten-free minor cereals and also of pseudocereals. The analysed samples showed a high content of bioactive compounds; in particular, amaranth, cañihua and quinoa are good sources of vitamin E, while millet, sorghum and teff (*Eragrostis tef*, or William's Lovegrass) are good sources of thiamine. Moreover, millet provides a fair amount of carotenoids, and in particular of lutein. These data can provide more information on bioactive compounds in gluten-free grains. The use of these grains can improve the nutritional quality of gluten-free cereal-based products, and could avoid the monotony of the celiac diet.

Keywords: minor cereal; pseudocereal; bioactive compound; gluten-free grain; tocols; carotenoids

1. Introduction

Celiac disease is a chronic systemic, autoimmune disorder in genetically-predisposed individuals, triggered by exposure to dietary gluten, and resulting in mucosal inflammation, villous atrophy and crypt hyperplasia [1]. It is characterised by an abnormal immune reaction consisting of an excessive response of the immune system to a group of cereal proteins, called prolamines (gliadin, hordein, sekalina, avenin), which are found in wheat, barley, rye and oats. Celiac disease affects approximately 1% of the world population, and it has significantly increased due to an underestimation, since it is often left undiagnosed [2]. The only treatment for people with the celiac problem is the adherence to gluten-free foods for their whole lifetime.

Several studies demonstrated that sticking to a gluten-free diet for a lifetime can lead to a nutritional imbalance in celiac subjects, such as a malabsorption of nutrients, and deficiencies of several vitamins and minerals. These deficiencies are due both to the phenomena of malabsorption at the intestinal level, and to the monotony of a diet based mainly on rice and maize [3–6].

Recently, more attention has been given to gluten-free minor cereals and pseudocereals as alternatives to those conventionally used for celiacs. Many of them have been defined as "orphan crops" or "underutilised crops"; they are indigenous crops scarcely documented and rarely used by food industries [7]. Many underutilised crops are relatively more drought-tolerant than most major cereals; they play a significant role in many developing countries, providing food security and income to resource-poor farmers [8].

Gluten-free alternative sources studied in this work include minor cereals (sorghum, teff, millet and wild rice), and pseudocereals (quinoa, cañihua, chia, and amaranth). These grains are mainly consumed as flours and seeds, which can be added to preparations such as soups, yogurt, cakes, breads and others cereal-based products; nevertheless, any commercialisation of these products is still quite limited in the Italian market. Some of these are a source of nutrients and bioactive compounds that could improve the nutritional quality of gluten-free products.

Carotenoids are a significant group of bioactive compounds with health promoting properties [9,10] and are responsible for the colour of a wide variety of grains [11]. Some carotenoids are the precursors of retinol (vitamin A), and are very strong natural antioxidants. Carotenoids are known to be efficient physical and chemical quenchers of singlet oxygen, as well as potent scavengers of other reactive oxygen species [9]. Vitamin E is a natural antioxidant comprising two groups of vitamers, tocopherols and tocotrienols, occurring in eight forms: α-tocopherol (α-T), β-tocopherol (β-T), γ-tocopherol (γ-T), and δ-tocopherol (δ-T) and α-tocotrienol (α-T3), β-tocotrienol (β-T3), γ-tocotrienol (γ-T3), and δ-tocotrienol (δ-T3). Vegetable oils are the main tocol sources, however, substantial amounts of these compounds are also reported in most cereal grains [12–14]. The potential health benefits of tocols include the prevention of certain types of cancer, heart diseases and other chronic diseases [15,16]. Thiamine (B1) is one of the major water-soluble vitamins, as it plays an important role as a co-factor of several key enzymes involved in the carbohydrate metabolism and defence mechanism [17]. It can be found in moderate amounts in all foods: Nuts and seeds, legumes, wholegrain/enriched cereals and breads, as well as pork [18]. Thiamine deficiency is rare in healthy individuals in food-secure settings, where access to thiamine-rich foods ensures adequate intakes [19]. Riboflavin (B2) is a precursor of the co-enzymes flavin mononucleotide (FMN; riboflavin phosphate) and flavin adenine dinucleotide (FAD), which are components of oxidases and dehydrogenases. It is also important for skin health and normal vision, and can be found in whole cereals, breads, leafy green vegetables and milk products [18].

To date, the evaluation of nutritional and bioactive compound profiles of alternative gluten-free grains is limited, if not lacking [20–23]. These researches are of a great importance in order to formulate gluten-free cereal-based products with a higher nutritional value. Thus, in this work, samples of minor cereals and pseudocereals commercialised in Italy have been characterised for their nutritional value, with a particular focus on some bioactive compounds, such as carotenoids, tocols, thiamine and riboflavin, in order to increase the awareness of their nutritional profile. Moreover, data coming from this study may be included in food nutrient databases.

2. Material and Methods

2.1. Sample Collection and Preparation

Thirty one different minor cereals and pseudocereals were bought in Italian specialised shops (Table 1). Different brands were considered for each grain. Grains were grounded with a refrigerated IKA A10 laboratory mill (Staufen, Germany), then carefully mixed and stored at −20 °C until analysis. Each sample was analysed in triplicate.

Table 1. List of analysed gluten-free grains.

Minor Cereals	Samples (*n*)
Millet (*Panicum miliaceum* L.)	6
White Sorghum (*Sorghum bicolor* L.)	3
Teff (*Eragrostis tef* (Zucc.) Trotter)	4
Wild rice (*Zizania aquatica* L.)	2

Table 1. *Cont.*

Pseudocereals	
White quinoa (*Chenopodium quinoa* Willd.)	3
Pigmented quinoas (red and black) (*Chenopodium quinoa* Willd.)	4
Cañihua (*Chenopodium pallidicaule*)	3
Amaranth (*Amaranthus* spp.)	3
Chia (*Salvia hispanica* L.)	3

2.2. Chemical Analysis

2.2.1. Proximate Analysis

Moisture, ash, fat, and protein contents were determined using an ICC standard procedure [24]. Briefly, moisture was determined using an oven set at 130 °C, and ash was quantified using a muffle furnace set at 525 °C. The protein content was determined though the Kjeldhal method (N × 6.25), and lipids were determined by the Soxhlet method. Carbohydrates plus fibre were calculated as a difference, using the following equation: (100 − (% moisture + % lipids + % proteins + % ash)).

2.2.2. Carotenoid Analysis

Carotenoid extraction was carried out using the saponification method reported by Panfili et al. [14]. About 0.2 g of milled sample was weighed and placed in a screw-capped tube. Then, 5 mL of ethanolic pyrogallol (60 g/L) was added as an antioxidant, followed by 2 ml of absolute ethanol, 2 mL of sodium chloride (10 g/L) and 2 mL of potassium hydroxide (600 g/L). The tubes were placed in a 70 °C water bath and mixed every 5–10 min during saponification. After alkaline digestion at 70 °C for 45 min, the tubes were cooled in an ice bath, and 15 mL of sodium chloride (10 g/L) were added. The suspension was then extracted twice with 15 mL portions of n-hexane/ethyl acetate (9:1, *v*/*v*). The organic layers, containing carotenoids, were collected and evaporated to dryness; the dry residue was dissolved in 2 mL of isopropyl alcohol (10%) in *n*-hexane. A HPLC Dionex (Sunnyvale, CA) analytical system, consisting of a U6000 pump system and a 50 µL injector loop (Rheodyne, Cotati) was used. The chromatographic separation of the compounds was achieved by means of a 250 mm × 4.6 mm i.d., 5 µm particle size, Kromasil Phenomenex Si column (Torrance, CA, USA). The mobile phase was *n*-hexane/isopropyl alcohol (5%) at a flow rate of 1.5 mL/min. Spectrophotometric detection was achieved by means of a diode array detector set in the range of 350–500 nm. Peaks were detected at 450 nm. Carotenoids were identified through their spectral characteristic, and comparison of their retention times with known standard solutions. Data were stored and processed by a Dionex Chromeleon Version 6.6 chromatography system (Sunnyvale, CA, USA). All-trans-β-carotene and lutein were obtained from Sigma Chemicals (St. Louis, MO, USA); zeaxanthin and β-cryptoxanthin were obtained from Extrasynthese (Z.I. Lyon-Nord, Genay, France).

2.2.3. Tocol Analysis

Tocols were determined after the same saponification method described for carotenoids. An aliquot of the carotenoid extract was collected and evaporated to dryness, and the dry residue was dissolved in 2 mL of isopropyl alcohol (1%) in *n*-hexane, and was analysed by HPLC, under normal phase conditions, using a 250 × 4.6 mm i.d., 5 mm particle size Kromasil Phenomenex Si column (Torrance, CA, USA) [14]. Fluorometric detection of all compounds was performed at an excitation wavelength of 290 nm and an emission wavelength of 330 nm by means of an RF 2000 spectrofluorimeter (Dionex, Sunnyvale, CA, USA). The mobile phase was *n*-hexane/ethylacetate/acetic acid (97.3:1.8:0.9 *v*/*v*/*v*), at a flow rate of 1.6 mL/min [14,25]. Compounds were identified by a comparison of their retention times with those of known available standard solutions, and quantified through the calibration curves of the standard solutions. The concentration range was 5–25 µg/mL for every tocol standard. Vitamin E activity was expressed as Tocopherol Equivalent (T.E.) (mg/100 g of fresh weight f.w.), calculated as reported by Sheppard et al. [26].

2.2.4. Thiamine and Riboflavin Analysis

Thiamine and riboflavin were extracted as in Hasselman et al. [27]. Briefly, samples were placed in 100 mL volumetric flasks containing 20 mL of 0.1 N HCl and heated in a water bath at 100 °C for 30 min. After cooling at room temperature, the pH of the samples was adjusted to 4.5 with 2.5 M NaOAc. Following the addition of 0.2 mL of aqueous Clara-Diastase (50 mg/mL), these samples were incubated for 3 h at 37 °C. After cooling, the samples were brought up to 25 mL with distilled water. Then these same samples were centrifuged and filtered through a 0.45 µm filter. Thiamine was converted to thiochrome by adding 1.25 mL of 1% potassium ferricyanide in 15% aqueous NaOH to 2.5 mL of filtered extract. After 1 min for oxidation, 0.25 mL of 85% H_3PO4 was added. The extract was purified on a Sep-Pak C18 cartridge. The cartridge was washed with 5 mL MeOH, followed by 5 ml of 0.05 M NH_4OAc (adjusted to pH 5.0 (acidic) with HOAc). The sample (5 mL) was loaded into a Sep-Pak C18 cartridge, and then the cartridge was washed with 0.05 M NH_4OAc and, finally, the vitamins were eluted with 5 mL mobile phase. Extracts were separated by a HPLC Dionex (Sunnyvale, CA, USA), with a U3000 pump and an injector loop (Rheodyne, Cotati). Separation was made at a flow rate of 0.8 mL/min with Methanol: NaOAc (40:60 *v/v*) as a mobile phase, by using a 5 µm C18 Luna, Phenomenex (Torrance, CA, USA) stainless steel column (250 × 4.6 mm i.d.). Fluorometric detection was performed at an excitation wavelength of 366 nm and an emission wavelength of 453 nm for thiamine, and an excitation wavelength of 453 nm and an emission wavelength of 580 nm for riboflavin, by means of an RF 2000 spectrofluorimeter (Dionex, Sunnyvale, CA, USA). Data were processed by a Dionex Chromeleon Version 6.6 chromatography system (Sunnyvale, CA, USA). Thiamine and riboflavin were compared with known available standards, and identified considering their retention times and relative elution order. Thiamine and riboflavin standards were obtained from Sigma Chemicals (St. Louis, MO, USA).

3. Results and Discussion

3.1. Nutritional Composition

The nutritional composition of analysed minor cereals and pseudocereals is shown below in Table 2.

Table 2. Nutritional composition of gluten-free grains (g/100 g).

	Minor Cereals				Pseudocereals			
	Millet	Sorghum	Teff	Wild Rice	Quinoa (White and Pigmented)	Cañihua	Amaranth	Chia
Moisture	12.7 (2.0) [a]	12.5 (6.9)	11.5 (1.4)	10.5 (0.4)	11.5 (9.8)	8.6 (5.7)	11.0 (1.0)	8.4 (6.7)
Ash	1.0 (63.0)	1.4 (8.6)	2.3 (5.6)	1.8 (7.2)	2.2 (3.0)	2.4 (6.6)	2.3 (8.7)	4.5 (2.7)
Protein	11.7 (3.3)	9.0 (0.1)	11.7 (1.7)	12.4 (6.1)	12.9 (1.5)	14.1 (2.6)	13.8 (3.4)	21.5 (7.6)
Fat	4.4 (0.4)	2.6 (26.9)	2.4 (4.1)	1.2 (4.5)	5.8 (12.0)	8.4 (1.1)	6.1 (5.7)	35.4 (2.1)
Carbohydrate + Fibre *	70.2	74.5	72.1	74.1	67.6	66.8	66.8	30.2

* Calculated by difference; [a]: coefficient of variability.

The composition of the chia seeds notably differs from all the other cereal and pseudocereal samples, showing high concentrations of fats (35.4 g/100 g), proteins (21.5 g/100 g) and ash (4.5 g/100 g). These values are similar to those observed by other authors for the chia seeds [28]. In general, wild rice and pseudocereals are a good source of protein. Taking European law into account [29], wild rice, all quinoa seeds, cañihua and amaranth can be declared in a label with the claim "source of protein", since they contain at least 12 g of protein per 100 g. Chia seeds can be declared with a "high protein content", since they contain at least 20 g of protein per 100 g. The fat content was significantly higher for pseudocereals, if compared to minor cereals. Wild rice shows the lower fat content (1.2 g/100 g).

3.2. Carotenoids

Table 3 shows the carotenoid amounts of analysed samples. Carotenoids content (μg/100 g dry weight d.w.) varied significantly from 22 μg/100 g in amaranth to 763 μg/100 g in millet. In all gluten-free grains the main compounds are lutein and zeaxanthin. A comparison with the literature related to the HPLC analysis of carotenoids is very difficult, since the available few data are obtained by different methods, and these pigments may vary depending on genotype and location. The total carotenoid content of millet, wild rice, quinoas and cañihua is comparable with that of wheat (about 305 μg/100 g for durum and about 150 μg/100 g for soft wheat) [12,30], and of pigmented rice (460–50 μg/100 g) [31], but it is significantly lower than that of maize (about 1110 μg/100 g) [30,32]. Among minor cereals, literature data are reported only for sorghum [33], where the authors found an average amount of 20 μg/100 g as the sum of lutein and zeaxanthin, with a high variability among the different genotypes.

Table 3. Carotenoid composition in gluten-free grains (μg/100 g d.w.).

Carotenoids	Minor Cereals			Pseudocereals					
	Millet	Sorghum	Teff	Wild Rice	White Quinoa	Pigmented Quinoas	Cañihua	Amaranth	Chia
β-Carotene	19.8 (15.0) [a]	9.86 (10.0)	7.8 (20.0)	6.23 (10.0)	12.3 (10.0)	23.6 (23.0)	20.2 (28.0)	tr	12.4 (10.0)
β-Criptoxanthin	20.0 (30.0)	nd	tr	tr	tr	tr	tr	nd	nd
Lutein	535.5 (3.4)	11.2 (64.0)	36.45 (30.0)	196.2 (36.6)	85.6 (1.3)	265.2 (33.0)	325.3 (0.1)	19.8 (5.0)	tr
Zeaxanthin	188.3 (10.0)	28.9 (10.0)	18.4 (40.0)	9.7 (10.0)	11.2 (11.0)	13.2 (30.0)	40.2 (4.2)	2.2 (11.3)	33.5 (10.0)
Total Carotenoid	763.1 (4.0)	50.46 (8.0)	62.6 (28.0)	212.3 (8.0)	109.1 (11.0)	302.0 (26.0)	385.7 (10.0)	22.0 (10.0)	45.9 (8.0)

[a]: Coefficient of variability; nd: not detectable; tr: traces.

In the present study, the variability of the total carotenoid content within the same cereal (expressed by the coefficient of variability, CV%), is from 4% in millet to 26% in pigmented quinoa. This variability may be due to genetic, pedoclimatic and varietal factors [34]. Regarding pseudocereals, results for chia are similar to those obtained in the work of da Silva et al. [28]. Significant differences between white and pigmented quinoas were found for total carotenoids, due to the different lutein amounts, as also observed by Tang et al. [35], who indicate a direct correlation between the higher total carotenoid content and the darkness of the seed coat.

3.3. Tocols

The characterisation of tocols in minor cereals and pseudocereals is reported in Table 4. Except for wild rice, which shows a minor content of total tocols (TC) (about 0.4 mg/100 g), the TC of minor cereals and amaranth are comparable with that of wheat, maize and rice (about 3.5–7.0, 6.0–7.0 and 2.3–2.7 mg/100g, respectively) [12,14,36] while, for the remaining pseudocereals, these values are significantly higher. Among minor cereals, teff shows the highest amount of total tocols (6 mg/100g d.w.), followed by millet and sorghum with about 4 and 3 mg/100g respectively.

Table 4. Tocol composition in gluten-free grains (mg/100g d.w.)

Tocols	Minor Cereals				Pseudocereals			
	Millet	Sorghum	Teff	Wild Rice	Quinoa (White and Pigmented)	Cañihua	Amaranth	Chia
α-T	0.16 (6.2) [a]	0.60 (83.0)	0.11 (18.2)	0.13 (11.5)	2.86 (9.58)	4.2 (35.7)	1.28 (44.5)	0.33 (33.3)
β-T	0.06 (16.6)	0.08 (62.5)	0.06 (20.0)	0.10 (13.0)	0.11 (23.0)	0.28 (21.0)	3.43 (46.0)	nd [b]
γ-T	2.73 (47.2)	2.32 (41.4)	5.52 (8.3)	0.10 (1.0)	5.9 (8.3)	12.50 (4.3)	0.30 (36.7)	13.59 (21.5)
δ-T	0.45 (29.0)	0.03 (33.0)	0.14 (14.0)	nd	0.22 (1.0)	0.40 (5.0)	1.28 (35.0)	0.38 (34.0)
α-T3	nd	nd	nd	nd	nd	nd	nd	nd
β-T3	0.12 (50.0)	nd	nd	0.03 (16.6)	tr	nd	nd	nd
γ-T3	0.04 (25.0)	nd	0.15 (73.0)	nd	tr	nd	nd	0.13 (23.0)
δ-T3	0.24 (45.8)	nd	nd	nd	nd	nd	nd	nd
Total tocols	3.80 (45.0)	3.09 (51.0)	5.99 (5.0)	0.36 (1.0)	9.10 (8.0)	18.06 (3.9)	6.31 (42.0)	14.43 (22.0)
T.E.	0.43 (28.0)	0.78 (82.0)	0.56 (21.0)	0.17 (5.9)	3.62 (1.0)	4.5 (2.0)	2.7 (33.0)	1.6 (24.0)

[a]: Coefficient of variability; [b]: Not detectable; tr: traces; T.E.: Tocopherol equivalent (mg/100g f.w.).

Except for wild rice, where α-tocopherol is the prevalent isomer, the main tocopherol isomer is γ-tocopherol, which represents the 92%, 72% and 75% of the total content in teff, millet and sorghum, respectively. For pseudocereals, the highest content of total tocols was found in cañihua (about 18 mg/100 g), followed by chia seeds (about 14 mg/100 g d.w.) and quinoas, with an average of 9.1 mg/100 g d.w. Contrarily to carotenoids, among all analysed quinoa seeds, all of the found vitamers did not show significant qualitative and quantitative differences. Amaranth is the pseudocereal with the lowest total tocol amounts (about 6 mg/100g). For chia, cañihua and quinoa the predominant isomer is γ-tocopherol (94%, 69% and 64% of total tocols), while for amaranth the prevalent isomer is β-tocopherol, which represents 54% of the total tocols.

γ-Tocopherol has also been found as the main vitamer in quinoa and chia in other works [28,35,37]. References for tocols are not available for all analysed gluten-free grains and, where present, they show similar results in millet and sorghum [3,23]. Moreover a comparison with the literature data related to tocol analysis is very difficult, for the same reasons already explained for carotenoids.

Table 4 also reports values of vitamin E activity provided by 100 g of product, expressed as Tocopherol Equivalent (T.E.) (mg/100 g product) [26]. Taking into account the Recommended Daily Allowance (RDA) for vitamin E, which is of 12 mg/day [38], 100 g of amaranth contribute to 22% of the RDA, while quinoas and cañihua approximately to 35% of the RDA, so as to be declared in a label as a "source of vitamin E". A portion of these pseudocereals (70 g) contributes approximately to 15% of the RDA for amaranth and to 25% of the RDA for quinoas and cañihua.

3.4. Thiamine and Riboflavin

Table 5 reports the values of the thiamine and riboflavin of analysed grains. The concentrations of thiamine are different between minor cereals and pseudocereals, except for wild rice. In whole wheat grains about 0.40 mg/100g are found in the literature [39,40]. Low values of riboflavin were found for all samples, except for wild rice, with values comparable to those of whole wheat grains and maize (0.15 and 0.20 mg/100g, respectively) [39,40].

Table 5. Thiamine and riboflavin content in gluten-free grains (mg/100g d.w.).

	Thiamine (mg/100g d.w.)	CV%	%RDA (1.1 mg/100g f.w.)	Riboflavin (mg/100g d.w.)	CV%	%RDA (1.4 mg/100g f.w.)
Minor Cereals						
Millet	0.28	49	23	0.02	25	1
Sorghum	0.28	61	23	tr	75	-
Teff	0.22	35	17	0.02	15	1
Wild rice	0.08	28	6	0.17	3	11
Pseudocereals						
Quinoa (white and pigmented)	0.13	50	9	0.02	32	1
Cañihua	0.04	6	3	0.09	8	6
Amaranth	0.03	23	3	0.01	20	1
Chia	0.06	2	5	0.02	20	1

tr: traces.

Taking into account the Recommended Daily Allowance (RDA) for thiamine, which is of 1.1 mg/day [38], 100 g of teff would contribute to approximately 17% of the RDA, while 100 g of millet and sorghum to 23% of the RDA, so as to be declared in a label as a "source of thiamine". A portion of 80 g contributes approximately to 16% of the RDA for teff and to 20% of the RDA for millet and sorghum.

4. Conclusion

Naturally gluten-free products are corn, rice, potatoes, soybean, millet, buckwheat, tapioca, amaranth, cassava, lentils, beans, sago, sorghum, nuts, as well as meat, fruit and vegetables. Among these, cereals and pseudocereals are becoming increasingly important. This work confirms that minor cereals and pseudocereals are an important source of bioactive compounds. In particular, wild rice and all analysed pseudocereals are good sources of protein. Taking into account the Recommended Daily Allowance (RDA) for vitamins established by the Commission of the European Communities, amaranth, cañihua and quinoa can be declared on the label as a source of vitamin E, the main antioxidant found in cells involved in the prevention of several diseases. Moreover, millet, sorghum and teff can be declared on the label as a potential source of thiamine. Millet also provides a fair amount of lutein. In the light of these results, it is possible to use the combined mix of these flours in order to improve the nutritional value of cereal-based gluten-free products.

Author Contributions: Conceptualization, writing-review and editing, G.P; investigation and writing-original draft preparation, S.N. and A.D.A.; data curation, writing-review and editing A.F; writing-review and editing, L.C.

Funding: This research received no external funding.

Conflicts of Interest: The authors declare no conflict of interest.

References

1. Nasr, I.; Nasr, I.; Al Shekeili, L.; Al Wahshi, H.A.; Nasr, M.H.; Ciclitira, P.J. Celiac disease, wheat allergy and non celiac gluten sensitivity. *Integr. Food Nutr. Metab.* **2016**, *3*, 336–340. [CrossRef]
2. Rubio-Tapia, A.; Kyle, R.A.; Kaplan, E.L.; Johnson, D.R.; Page, W.; Erdtmann, F.; Brantner, T.L.; Kim, W.R.; Phelps, T.K.; Lahr, B.D.; et al. Increased prevalence and mortality in undiagnosed celiac disease. *Gastroenterology* **2009**, *137*, 88–93. [CrossRef] [PubMed]
3. Saturni, L.; Ferretti, G.; Bacchetti, T. The gluten-free diet: Safety and nutritional quality. *Nutrients* **2010**, *2*, 16. [CrossRef] [PubMed]
4. Vici, G.; Belli, L.; Biondi, M.; Polzonetti, V. Gluten free diet and nutrient deficiencies: A review. *Clin. Nutr.* **2016**, *35*, 1236–1241. [CrossRef] [PubMed]
5. Gobbetti, M.; Pontonio, E.; Filannino, P.; Rizzello, C.G.; De Angelis, M.; Di Cagno, R. How to improve the gluten-free diet: The state of the art from a food science perspective. *Food Res. Int.* **2018**, *110*, 22–32. [CrossRef] [PubMed]

6. De Lourdes Moreno, M.; Comino, I.; Sousa, C. Update on nutritional aspects of gluten-free diet in celiac patients. *J. Nutr.* **2014**, *1*, 7–18.
7. Cheng, A.; Mayes, S.; Dalle, G.; Demissew, S.; Massawe, F. Diversifying crops for food and nutrition security-a case of teff. *Biol. Rev.* **2017**, *92*, 188–198. [CrossRef]
8. Cheng, A. Review: Shaping a sustainable food future by rediscovering long-forgotten ancient grains. *Plant Sci.* **2018**, *269*, 136–142. [CrossRef]
9. Fiedor, J.; Burda, K. Potential role of carotenoids as antioxidants in human health and disease. *Nutrients* **2014**, *6*, 466. [CrossRef]
10. Eggersdorfer, M.; Wyss, A. Carotenoids in human nutrition and health. *Arch. Biochem. Biophys.* **2018**, *652*, 18–26. [CrossRef]
11. Fratianni, A.; Mignogna, R.; Niro, S.; Panfili, G. Determination of lutein from fruit and vegetables through an alkaline hydrolysis extraction method and HPLC analysis. *J. Food Sci.* **2015**, *80*, 2686–2691. [CrossRef] [PubMed]
12. Fratianni, A.; Giuzio, L.; Di Criscio, T.; Zina, F.; Panfili, G. Response of carotenoids and tocols of durum wheat in relation to water stress and sulfur fertilization. *J. Agric. Food Chem.* **2013**, *61*, 2583–2590. [CrossRef] [PubMed]
13. Mignogna, R.; Fratianni, A.; Niro, S.; Panfili, G. Tocopherol and tocotrienol analysis as a tool to discriminate different fat ingredients in bakery products. *Food Control.* **2015**, *54*, 31–38. [CrossRef]
14. Panfili, G.; Fratianni, A.; Irano, M. Normal phase high-performance liquid chromatography method for the determination of tocopherols and tocotrienols in cereals. *J. Agric. Food Chem.* **2003**, *51*, 3940–3944. [CrossRef] [PubMed]
15. Tiwari, U.; Cummins, E. Nutritional importance and effect of processing on tocols in cereals. *Trends Food Sci. Technol.* **2009**, *20*, 511–520. [CrossRef]
16. Shahidi, F.; de Camargo, A.C. Tocopherols and tocotrienols in common and emerging dietary sources: Occurrence, applications, and health benefits. *Int. J. Mol. Sci.* **2016**, *17*, 1745. [CrossRef] [PubMed]
17. Martin, P.R.; Singleton, C.K.; Hiller-Sturmhöfel, S. The role of thiamine deficiency in alcoholic brain disease. *Alcohol. Res. Health* **2003**, *27*, 134–142. [PubMed]
18. Zhang, Y.; Zhou, W.; Yan, J.; Liu, M.; Zhou, Y.; Shen, X.; Ma, Y.; Feng, X.; Yang, J.; Li, G. A review of the extraction and determination methods of thirteen essential vitamins to the human body: An update from 2010. *Molecules* **2018**, *23*, 1484. [CrossRef] [PubMed]
19. Thurnham, D.I. Thiamin: Physiology. *Encycl. Hum. Nutr.* **2013**, *4*, 274–279.
20. Vilacundo, R.; Hernández-Ledesma, B. Nutritional and biological value of quinoa (*Chenopodium quinoa* Willd). *Curr. Opin. Food Sci.* **2017**, *14*, 1–6. [CrossRef]
21. Zhu, F. Chemical composition and food uses of teff (Eragrostis tef). *Food Chem.* **2018**, *239*, 402–415. [CrossRef] [PubMed]
22. Alvarez-Jubete, L.; Arendt, E.K.; Gallagher, E. Nutritive value of pseudocereals and their increasing use as functional gluten free ingredients. *Trends Food Sci. Technol.* **2010**, *21*, 106–113. [CrossRef]
23. Asharani, V.T.; Jayadeep, A.; Malleshi, N.G. Natural antioxidants in edible flours of selected small millets. *Int. J. Food Prop.* **2010**, *13*, 41–50. [CrossRef]
24. ICC. *International Association for Cereal Science and Technology*; ICC: Vienna, Austria, 1995.
25. Fratianni, A.; Caboni, M.F.; Irano, M.; Panfili, G. A critical comparison between traditional methods and supercritical carbon dioxide extraction for the determination of tocochromanols in cereals. *Eur. Food Res. Technol.* **2002**, *215*, 353–358. [CrossRef]
26. Sheppard, A.J.; Pennington, J.A.T.; Weihrauch, J.L. Analysis and distribution of vitamin E in vegetable oil and foods. In *Vitamin E in Health and Disease*; Packer, L., Fuchs, J., Eds.; Marcel—Dekker: New York, NY, USA, 1993.
27. Hasselmann, C.; Franck, D.; Grimm, P.; Diop, P.A.; Soulrs, C. High performance liquid chromatographic analysis of thiamin and riboflavin in dietetic foods. *J. Micronutr. Anal.* **1989**, *5*, 269–279.
28. Da Silva, B.P.; Anunciação, P.C.; Matyelka, J.C.D.S.; Della Lucia, C.M.; Martino, H.S.D.; Pinheiro-Sant'Ana, H.M. Chemical composition of Brazilian chia seeds grown in different places. *Food Chem.* **2017**, *221*, 1709–1716. [CrossRef] [PubMed]
29. Regulation, E.C. No 1924/2006 of the European Parliament and of the Council of 20 December 2006 on nutrition and health claims made on foods. *Off. J. Eur. Union* **2007**, *12*, 3–18.

30. Panfili, G.; Fratianni, A.; Irano, M. Improved normal-phase high-performance liquid chromatography procedure for the determination of carotenoids in cereals. *J. Agric. Food Chem.* **2004**, *52*, 6373–6377. [CrossRef]
31. Melini, V.; Panfili, G.; Fratianni, A.; Acquistucci, R. Bioactive compounds in rice on Italian market: Pigmented varieties as a source of carotenoids, total phenolic compounds and anthocyanins, before and after cooking. *Food Chem.* **2019**, *277*, 119–127. [CrossRef]
32. Alfieri, M.; Hidalgo, A.; Berardo, N.; Redaelli, R. Carotenoid composition and heterotic effect in selected Italian maize Germplasm. *J. Cereal Sci.* **2014**, *59*, 181–188. [CrossRef]
33. De Morais Cardoso, L.; Pinheiro, S.S.; Da Silva, L.L.; de Menezes, C.B.; de Carvalho, C.W.P.; Tardin, F.D.; Queiroz, V.A.V.; Martino, H.S.D.; Pinheiro-Sant'Ana, H.M. Tocochromanols and carotenoids in sorghum (*Sorghum bicolor* L.): Diversity and stability to the heat treatment. *Food Chem.* **2015**, *172*, 900–908. [CrossRef] [PubMed]
34. Abdel-Aal, E.; Wood, P. *Specialty Grains for Food and Feed*; American Association of Cereal Chemists: Paul, MN, USA, 2005.
35. Tang, Y.; Li, X.; Chen, P.X.; Zhang, B.; Hernandez, M.; Zhang, H.; Marcone, M.F.; Liu, R.; Tsao, R. Characterisation of fatty acid, carotenoid, tocopherol/tocotrienol compositions and antioxidant activities in seeds of three Chenopodium quinoa Willd. genotypes. *Food Chem.* **2015**, *174*, 502–508. [CrossRef] [PubMed]
36. Irakli, M.N.; Samanidou, V.F.; Katsantonis, D.N.; Biliaderis, C.G.; Papadoyannis, I.N. Phytochemical profiles and antioxidant capacity of pigmented and non-pigmented genotypes of rice (*Oryza sativa* L.). *Cereal Res. Commun.* **2016**, *44*, 98–110. [CrossRef]
37. Pereira, E.; Encina-Zeladaa, C.; Barrosa, L.; Gonzales-Barrona, U.; Cadaveza, V.; Ferreira, I.C.F.R. Chemical and nutritional characterization of Chenopodium quinoa Willd (quinoa) grains: A good alternative to nutritious food. *Food Chem.* **2019**, *280*, 110–114. [CrossRef] [PubMed]
38. Regulation, E.U. No 1169/2011 of the European Parliament and of the Council of 25 October 2011 on the provision of food information to consumers. *Off. J. Eur. Union* **2011**, *50*, 18–63.
39. Călinoiu, L.F.; Vodnar, D.C. Whole grains and phenolic acids: A review on bioactivity, functionality, health benefits and bioavailability. *Nutrients* **2018**, *10*, 1615. [CrossRef] [PubMed]
40. Gwirtz, J.A.; Garcia-Casal, M.N. Processing maize flour and corn meal food products. *Ann. N. Y. Acad. Sci.* **2014**, *1312*, 66–75. [CrossRef]

© 2019 by the authors. Licensee MDPI, Basel, Switzerland. This article is an open access article distributed under the terms and conditions of the Creative Commons Attribution (CC BY) license (http://creativecommons.org/licenses/by/4.0/).

foods

MDPI

Article

Micronutrient Analysis of Gluten-Free Products: Their Low Content Is Not Involved in Gluten-Free Diet Imbalance in a Cohort of Celiac Children and Adolescent

Idoia Larretxi [1,2], Itziar Txurruka [1,2], Virginia Navarro [1,2], Arrate Lasa [1,2], María Ángeles Bustamante [1,*], María del Pilar Fernández-Gil [1], Edurne Simón [1,2] and Jonatan Miranda [1,2]

1 Gluten Analysis Laboratory of the University of the Basque Country, Department of Nutrition and Food Science, University of the Basque Country, UPV/EHU, 01006 Vitoria, Spain

2 GLUTEN3S research group, Department of Nutrition and Food Science, University of the Basque Country, UPV/EHU, 01006 Vitoria, Spain

* Correspondence: marian.bustamante@ehu.eus; Tel.: +34-945-014086; Fax: +34-945-013014

Received: 19 July 2019; Accepted: 5 August 2019; Published: 7 August 2019

Abstract: Data about the nutritional composition of gluten-free products (GFP) are still limited. Most studies are based on ingredient and nutrition information described on the food label. However, analytical determination is considered the gold standard for compositional analysis of food. Micronutrient analytical content differences were observed in a selection of GF breads, flakes and pasta, when compared with their respective gluten-containing counterparts. In general terms, lower iron, piridoxin, riboflavin, thiamin, niacin, folate, manganese and vitamin B5 can be underlined. Variations in biotin and vitamin E content differed among groups. In order to clarify the potential contribution of the GFP to the gluten-free diet's (GFD) micronutrient shortages, analytical data were used to evaluate GFD in a cohort of celiac children and adolescent. Participants did not reach recommendations for vitamin A, vitamin E, folic acid, vitamin D, biotin, iodine, and copper. It does not seem that the lower micronutrient content of the analyzed GFP groups contributed to the micronutrient deficits detected in GFD in this cohort, whose diet was not balanced. Nevertheless, GFP fortification for folate and biotin is proposed to prevent the deficiencies observed in GFD, at least in the case of pediatric celiac disease.

Keywords: celiac disease; gluten-free diet; gluten-free product; micronutrient; vitamin and minerals; dietary recommendation

1. Introduction

Celiac disease (CD) is a chronic immune-mediated inflammatory pathology triggered by the gluten in the diet of genetically predisposed individuals. The need to avoid this protein in the diet of celiac people brought about some years ago the development of specific cereal-based gluten-free products (GFP). Despite the fact that these GFP allowed them to include a wide variety of foods in their diets, in recent years researchers have highlighted differences in the nutrient composition of GFP with respect to gluten containing counterparts [1,2], leading to a minor health rating in some food-groups [3,4].

It is important to note that most of the studies about the nutrient composition of the GFP are based on ingredients and the nutrition information described on the food label [2–4]. To improve these data, some works, such as that carried out by Mazzeo et al. (2015) [5], take advantage of the retention factors for each nutrient, including losses due to heating or other food preparation steps. However, analytical determination is considered the gold standard for composition analysis of food. Accurate analysis could also provide detailed information about vitamins and minerals, which is not totally or commonly available on label [6]. Therefore, access to micronutrient data is already restricted to hardly any research [7–9].

Furthermore, a gluten-free diet (GFD) often implies some nutritional imbalances, as recognized in the literature [10,11]. Not only have inadequate fat, protein, sugar and fiber consumption been observed in GFD, but also a poor intake of micronutrients such as iron, zinc, magnesium, calcium, folate, vitamin D and B_{12} [12]. Similarly, celiac people seem to have lower blood values for hemoglobin, ferritin, vitamin D, and copper than the rest of the population [13,14]. There has been speculation about whether the characteristic composition of GFP is responsible for GFD inadequacy. A potential correlation between both facts has been proposed by others [15].

In the case of GFP, the use of raw material such as unenriched rice or maize refined flours, gums or enzymes in their formulation could lead to a different composition compared to their gluten containing homologues [16]. Moreover, as the micronutrient content of gluten-free pseudocereals and legumes is higher than that of the gluten free cereals [15,17], some authors proposed to promote their use in GFP formulation [12,18,19].

In view of the above, the aim of this study was to assess analytically the macronutrient and micronutrient content of a selection of GF breads, flakes and pasta, and to compare it with their respective gluten-containing counterparts. Additionally, in order to clarify the potential contribution of the GFP to the GFD's micronutrient shortages, vitamin and mineral analytical data were used to evaluate GFD in a cohort of celiac children and adolescents.

2. Materials and Methods

2.1. Analytical Nutrient Content of GF Bread, Breakfast Cereals and Pasta

The measured samples were thirty-seven selected GFP signed with the Crossed Grain symbol: 13 breakfast cereals, 12 breads and 12 pasta products (Supplementary Table S1). All the food items were purchased from the local market (Vitoria, Spain) and they were stored frozen (−20 °C) until analyzed. The analytically determined composition of GF foodstuffs was compared with the data of equivalent gluten-containing breads (n = 19), breakfast cereals (n = 18), and pasta products (n = 8), analyzed in the same way and at the same time for macronutrients, and with micronutrient data obtained from the Spanish Food Composition Database—BEDCA database [20]. These results were also compared with the data described in the food label of GFPs.

Analysis of the nutrient content of foodstuff has been carried out using official methods. Crude protein content was determined by the Kjeldahl method (AOAC, 960.52A) [21] in a Foss Kjeltec™ distillation unit (Höganäs, Sweden). Fat content was analyzed by the Soxhlet extraction method based on the official method (AOAC, 2003.05) [21], using a Soxtherm extraction system (Gerhardt, Bonn, Germany). Determinations were performed in duplicate.

For mineral determination, microwave-assisted digestion was carried out in a closed microwave device Mars 5 (CEM, Vertex, Barcelona, Spain) equipped with 8–24 teflon vessels and temperature controllers. The quantitative analysis of selenium, manganese and cooper was performed by using ICP-MS (7700x, Agilent Technologies, Palo Alto, CA, USA) and MicroMist micro-uptake glass concentric nebulizer (Glass Expansion, West Melbourne, Victoria, Australia). ICP-OES (Horiba Jobin Yvon Activa, Kyoto, Japan) was used with a quartz Meinhard concentric nebulizer, a Scott-type spray chamber and a standard quartz sheath connection between the spray chamber and the torch in the case of calcium, sodium, zinc and iron quantification. Working standard solutions of Ca, Na (0–20 mg/L), Fe, Zn and Se (0–100 µg/L) were prepared immediately prior to their use, by stepwise dilution of certified standard multi-element solution (100 mg/L) (Merck, Darmstadt, Germany) with HNO3 1.0 % *v*/*v* (Merck, Darmstadt, Germany). Additionally, a 10 mg/L multi-element standard solution (Y, Rh) from Inorganic Ventures (Equilab, Madrid, Spain) was also used as the internal standard in direct ICP-MS analysis.

As a step prior to vitamin quantification, samples were extracted by liquid-liquid extraction using an aqueous acidic mixture, centrifuged and filtrated, except for vitamin E. Biotin, Folate, Niacin, Pyridoxine, Riboflavin, Thiamine, vitamin B_5 and B_{12} were measured by liquid chromatography (LC) with triple quadrupole mass spectrometry detection. High purity (>95%) standards (Merck, Darmstadt, Germany) were used for the identification of each vitamin by positive ionization of the electrospray and multiple reaction monitoring. Quantification was developed using the standard addition method. Vitamin E determination was carried out by previous saponification of the samples, followed by a liquid-liquid extraction and purification of the extracts. Afterwards, high performance LC with the fluorescence detector method was used to analyze vitamin E in each extract. Quantification was performed by an external calibration method using the calibration curve of the tocopherol standard (Merck, Darmstadt, Germany). Analytical determinations of micronutrients were carried out once in each sample, but it was verified before the analysis that the reproducibility of the methods was less than 5%.

As mentioned, the micronutrient content of GF foodstuffs was compared with that of gluten-containing counterparts, obtained from the Spanish Food Composition Database—BEDCA database- [20]. Data for biotin in all studied food groups and copper in cereals were obtained from McCance and Widdowson's "composition of foods integrated dataset" from the United Kingdom [22]. No available data were found with regard to the manganese content of cereal flakes in food composition databases from the UK, Australia, the USA or Spain [20,22–24].

2.2. Dietary Assessment: Participants and Procedure

Eighty-three minor celiac (age: 3 to 18 years; 53 girls and 30 boys) from the Basque Country took part in the study. The age of the participants was selected due to their higher consumption of GFP compared to adults [25,26]. All participants received oral and written information about the nature and purpose of the survey, and all of them gave written consent for involvement in the study. This study was approved by the Ethical Committee in University of Basque Country (CEISH/76/2011 and CEISH/194M/2013).

The dietary assessment followed in the research was described elsewhere [26]: three days food records (two weekdays and one weekend day) were selected for each patient, 24-h food recalls (24HRs) were filled in by each celiac patient. Micronutrient intake was calculated by a computerized nutrition program system (AyS, Software, Tandem Innova, Inc., Huesca, Spain). The analytically measured vitamin and mineral content of tested GF products was added into the food composition database of the program before calculations. Dietary reference intakes (DRI) for the Spanish population issued by the Spanish Societies of Nutrition, Feeding and Dietetics (FESNAD) in 2010 were taken as references for the interpretation of the 24HRs [27].

2.3. Statistical Analysis

Results are presented as mean ± standard deviation (SD) of the mean. Statistical analysis was performed using SPSS 24.0 (SPSS Inc., Chicago, IL, USA). After confirming the normal distribution of lipid and protein content variables using Shapiro-Wilks normality, paired-samples student's t test was used for comparison. Due to their skewed distribution, micronutrients variables for analytical and database information were analyzed by Mann–Whitney *U*. The level of significance was set to $p < 0.05$.

3. Results and Discussion

3.1. Macronutrient Content of GF Rendered Foods

With the aim of assessing representative products of a GFD, GFP from the three main cereal food-types contributing to a balanced diet, such as flakes, pasta and bread, were selected. Protein and lipid contents of the three GFP groups analyzed are shown in Table 1. Results were compared to the nutritional composition of their gluten-containing counterparts. With regard to breads, lipid content was higher and the protein content was lower than that of gluten containing products. Similarly, GF bread has been described as poor in proteins and rich in fat content by others [28]. GF pasta provided a lower protein amount, although the comparison to gluten containing pasta did not reach statistical significance. In general terms, lower protein content in GFP than in their counterpart has been proposed by previous research [2–4]. Nevertheless, and in good accordance with our data, Missbach et al. did not observe this pattern in flakes [2].

Table 1. Analytical protein and lipid content in gluten-free rendered foodstuffs divided by food groups, compared to gluten-containing products, expressed by 100 g of foodstuffs.

	Cereal Flakes			Bread			Pasta		
	GFP	GCP	*P*	GFP	GCP	*p*	GFP	GCP	*p*
Lipids	3.9 ± 5.3	2.6 ± 2.0	NS	5.6 ± 4.2	3.5 ± 4.1	0.05	3.2 ± 4.5	2.2 ± 1.1	NS
Proteins	7.4 ± 0.7	7.8 ± 3.1	NS	2.4 ± 2.0	9.0 ± 1.5	<0.001	6.5 ± 1.4	9.8 ± 4.2	NS

Values are means ± SD. SD, standard deviation; GFP, gluten-free product; GCP, gluten-containing product; *p*, statistical significance; NS, not significant.

Some clues for justifying the results could be extracted from the list of ingredients of GFP (Supplementary Table S1). Rice and maize flours are extensively used in GFP, especially in breads, and according to composition databases, their protein content is lower than that of wheat. Moreover, maize and rice starches, usually added as a substitute, are especially poor in this macronutrient. For pasta and flakes, other ingredients could hinder the protein deficit, such as cocoa or eggs, soy protein or meat from the filled pasta. For lipids, the use of additives like mono and diglycerides of fatty acids (E-471) in GFP, especially in breads, could affect the final composition. However, this study did not consider the label information of ingredients of GCP, thus making conclusive statements is not possible.

It is important to point out that the comparative study between GFP and their homologues with gluten in the present work was performed as suggested by Staudacher and Gibson [6], by direct analytical methods and in paired form. As stated in the introduction, most of the studies evaluating the differences between both foodstuffs are based on nutrition information taken from the food label. For this reason, the analytical results obtained were compared to those reported in the nutritional panel information and some interesting data were collected. With regard to bread, experimental data reported a lower lipid (23%, $p = 0.07$) and higher protein (37%; $p = 0.03$) content than that supplied by the label. Similarly, in the case of cereal flakes, the measured protein amount was higher (19%; $p = 0.04$). No differences were observed between analyzed and labelled data in GF pasta.

In view of Regulation (EU) No 1169/2011 [29], the declared values on labels shall be average values based on (a) the manufacturer's analysis of the food; (b) a calculation from the known or actual average values of the ingredients used; (c) a calculation from generally established and accepted data. It is not possible for us to determine how each manufacturer calculated label information. However, it must be highlighted that nutrient variations observed in bread types are not within the tolerance ranges between label information and our direct food analysis (tolerance ranges: ±1.5 g for lipids and ±2 g for proteins, when its content in food is <10 g per 100 g). This information brings to light that previous studies about bread described in the literature could be reconsidered, and additionally, it validates, in part, others about pasta and cereals.

3.2. Micronutrient Content of GFP, Compared to Gluten-Containing Products

Despite the growing market of the GFP [30], data about their vitamin and mineral contribution remain scarce. Moreover, the data found in the literature are usually calculated from ingredients and their composition databases, which has been proposed to lead to overestimation [5]. Table 2 shows analytical micronutrient content of GF bread, flakes and pasta, compared to that of their gluten-containing counterparts. Lower iron, piridoxin, riboflavin and thiamin content was found in the three GFP groups analyzed. Niacin reduction was observed in GF flakes and breads. With regard to iron, similar results were found by Rybicka [8], who described that 273 of 408 GFP analyzed fulfilled less than 10% of recommended nutrient intake per portion and only 23 products were major contributors to daily intake (over 25% of recommendation intake per portion). In a study performed with 368 GFP, including flours, breads, pasta and cold cereals, overall it was observed that these kinds of products contained lower amounts of thiamin, riboflavin and niacin than the wheat product they were intended to replace [31]. These results are in line with the results obtained in the present study.

Folate content was lower in GF flakes and pasta types; manganese amount was lower only in GF pasta, and that of vitamin B_5 in GF flakes. As stated before, commonly used ingredients for GFP are maize and rice flours as well as a variety of starches (potato, corn), among others. It seems that removal of protein-rich fractions from flours may result in dramatic depletion of folates. Additionally, rice flours are not very rich in this vitamin [9]. In fact, we calculated a reduction of almost 80% of folate content in rice flour with respect to wheat flour ($p = 0.05$) comparing the nutrient composition of both flours obtained from food composition databases from the UK, Australia, the USA or Spain [20,22–24].

Several studies have claimed lower zinc and copper and higher sodium content for GFP [4,32]. However, no significant differences in those minerals were found in our data.

Finally, biotin content differed widely among groups, being higher in cereal flakes and lower in pasta GFP than in their counterparts. Moreover, we found that some GF cereals were fortified with biotin, thus explaining its higher content in this GF food group. Similarly, although vitamin E contribution from GFP was lower in flakes, no differences were observed in pasta and bread. Moreover, it is worth mentioning that half of the analyzed bread types showed a formulation with sunflower oil (Supplementary Table S1), which led to higher vitamin E content in those specific stuffs.

It is important to point out that food technology interventions to improve the shelf life and rheological properties of GFP have influenced their nutritional profile [12]. In order to avoid the absence of the mentioned micronutrients without fortifying foodstuffs, different strategies can be proposed: avoiding starch as a major ingredient, sourdough fermentation, and using less popular grain GF flour such as that from pseudocereals (buckwheat, quinoa, amaranth and teff) or legumes, including wholemeal forms of gluten-free cereals [18,19,33,34]. In our samples, only one out of twelve foodstuffs analyzed in each group contained pseudocereals in their ingredients list (4 to 5 g in 100 g), reflecting the need of more research on the properties and technological characteristics of these raw materials, and promotion of their use.

3.3. Micronutrient Intake in Celiac Children and Adolescents

It is known that GFD can lead to imbalanced macronutrient distribution. Our previous work [26] reported that celiac children and adolescents consumed more fat and less carbohydrate than recommended and pointed at GF rendered foods as one of the culprits. Thus, taking into account directly analyzed micronutrient content, their intake on that pediatric cohort was calculated considering their age group and gender, and compared to FESNAD recommendations (Supplementary Figure S1).

More than 1/4 of participants did not reach recommendations for vitamin A and vitamin E. Four out of ten children and adolescents with CD showed low intake of folic acid, which was even less than 66% of the recommendation for 25% of participants. Sixty percent of participants did not get that for vitamin D, and moreover, about 40% of them did not reach 25% of the recommendation. Most participants showed very low intakes of biotin, iodine and copper. Slightly over half the participants did not fulfil 50% of iodine recommendation and more than 40% were not able to achieve 25% of that of biotine. The intake of the rest of micronutrient was appropriate. With the exception of vitamin D, the results obtained differ from those obtained in similar pediatric research on celiac children, where low intake of iron, calcium, selenium and magnesium was observed [10,35,36].

Table 2. Analytical [4] micronutrient content in gluten-free rendered foodstuffs divided by food groups, compared to gluten-containing products, expressed by 100 g of foodstuffs.

Micronutients/Products	Cereal Flakes					Breads					Pasta				
	GCP		GFP		*p*	GCP		GFP		*p*	GCP		GFP		*p*
	Mean	SD	Mean	SD		Mean	SD	Mean	SD		Mean	SD	Mean	SD	
Calcium (mg)	141	183	22.3	16.8	NS	60.5	34.4	90.8	57.5	NS	22.9	10.0	27.3	27.3	NS
Iron (mg)	9.87	5.19	1.8	1.8	<0.001	6.75	19.99	1.1	0.9	0.009	1.83	0.88	0.7	0.5	0.002
Sodium (mg)	332	332	357.1	313.6	NS	423	280	570.8	248.1	NS	61.0	155	34.8	65.6	NS
Zinc (mg)	5.94	12.9	0.9	0.5	NS	0.84	0.48	0.5	0.4	NS	1.45	1.10	1.1	0.5	NS
Copper (mg)	0.2	0.2	0.3	0.3	NS	0.14	0.11	0.1	0.1	NS	0.31	0.20	0.1	0.1	NS
Manganese (mg)			0.4	0.4		0.58	0.51	0.2	0.3	NS	2.22	1.18	0.5	0.8	0.042
Biotin (ug)	2.3	2.9	40.7	95.7	0.001	12.7	13.6	9.74	21.6	NS	15.8	14.8	1.10	1.00	0.002
Folate (ug)	275	35.4	55.4	88.9	0.062	32.55	14.79	32.8	56.8	NS	19.4	10.3	3.14	3.42	0.003
Niacin (mg)	16.2	8.25	3.02	2.41	<0.001	3.49	2.15	1.02	1.46	0.004	3.75	3.46	3.62	10.58	NS
Piridoxin (mg)	1.80	0.90	1.19	4.15	0.001	0.12	0.10	0.02	0.02	<0.001	0.12	0.07	0.01	0.01	<0.001
Riboflavin (mg)	1.45	0.72	0.12	0.17	<0.001	0.13	0.11	0.04	0.09	0.001	0.07	0.04	0.01	0.02	0.002
Tiamin (mg)	1.32	0.70	0.19	0.28	<0.001	0.20	0.12	0.01	0.01	<0.001	0.19	0.15	0.03	0.04	0.001
B_5 (mg)	7.55	3.46	1.03	1.81	0.045	0.39	0.07	0.42	0.57	NS	0.70	0.40	0.29	0.36	0.067
B_{12} (ug)	0.85	0.52	3.68	4.18	0.03	0.02	0.05	88.2	236	0.026	0.04	0.08	0.73	1.37	NS
Vitamin E (mg)	3.00	6.22	0.14	0.47	0.01	0.30	0.33	1.04	1.2	NS	0.09	0.12	0.2	0.1	NS

Values are means ± SD. SD, standard deviation; GFP, gluten-free product; GCP, gluten-containing product; *p*, statistical significance; NS, not significant.

Considering all the above mentioned, it does not seem that the GFP groups analyzed contribute to the micronutrient deficits detected in young celiac people's diets. In fact, cereals have only a modest role as source of these micronutrients. It is important to highlight that in our previous study [26] we reported unhealthy dietary habits in these celiac children and adolescents: very low cereal and vegetable consumption, low fruit and nut intake and excessive meat consumption. Thus, general recommendations to promote healthy GFD should be given to amend the observed wrong habits. It is worth mentioning that this conclusion refers to our cohort, and that in other dietary patterns, GFPs role could be different.

It must be pointed out that, in the case of folic acid, we observed a lower content of this vitamin in GFP than in their gluten containing equivalents. In this regard, in Canada and USA [37,38] the fortification of wheat flour with folic acid is mandatory, but not for other alternative flours, such as the ones used in GFP. Taking into account the folate deficiency observed in GFD, its fortification in GFP or ingredients could be of interest for celiac children. Folate fortification measures could also be extended to biotin, whose widespread diet-deficiency in celiac population was alarming. In fact, some of the GF cereals analyzed were supplemented with this vitamin (Supplementary Table S1).

It is of interest to point out that some deficiency diseases found in celiac people, such as anemia, low bone density or zinc depletion [39] are not only justified by nutritional shortages. Other pathological situations such as systemic inflammation or intestinal microbiota alteration appear to contribute to the persistence of those deficiencies in some celiac individuals [12,40,41].

It has to be highlighted that this paper presents wide-ranging high-quality nutritional information about GF bread, pasta and cereal micronutrient content. This remains limited in the literature and even more so in food panels or in databases used for GFD design and evaluation, where it is crucial. Moreover, it has assessed not only GFP composition but also its dietetic role, discussing, in general terms, its involvement in micronutrient deficiencies of the GFD of children and adolescents. Nevertheless, extrapolation to celiac adults is limited and needs further research. Moreover, as proposed elsewhere [42], the bioavailability of GFP is a matter of concern that should also be taken into account in further studies. Finally, it is also of great interest to analyze the nutritional composition of GFPs considering their ingredients list to define the role of ingredients such as gluten free cereals or pseudocereals, starches and additives in the final composition of the product.

The practical outcomes of the present study are relevant in improving the universal guidelines for food fortification in CD [43,44]. Some individualized supplementation is usually proposed for celiac people based on micronutrient related blood monitoring. Nevertheless, GFP fortification for folate and biotin could contribute to preventing the deficiencies observed in GFD, at least in the case of celiac children and youngsters.

4. Conclusions

Even if lower micronutrient content was found in the analyzed GFP groups, this fact was not related with the micronutrient deficits detected in GFD in a cohort of celiac children and adolescent. Nevertheless, according to the obtained results, GFP fortification for folate and biotin seems to be a suitable proposal in order to prevent the deficiencies observed in GFD.

Supplementary Materials: The following are available online at http://www.mdpi.com/2304-8158/8/8/321/s1, Figure S1: Percentage of celiac children and adolescents who accomplished or did not achieve 2/3 of the dietary reference intake of the vitamins and minerals (proposed by the Federation of Spanish Societies of Nutrition and Dietetics, FESNAD), Table S1: Analyzed gluten free products and ingredients declared on the package label.

Author Contributions: E.S. carried out the experimental design. I.L. and I.T. analyzed lipid, protein and micronutrient content. V.N. and A.L. performed the analysis of diet. M.Á.B., M.d.P.F.-G. and J.M. analyzed all data and contributed to statistical analysis. I.L., I.T., V.N. and J.M. wrote the manuscript.

Funding: Idoia Larretxi is a fellowship of the University of the Basque Country, UPV/EHU. This research was supported by a grant from the Basque Government (Proyectos de Investigación Focalizada Agricultura PA15/01) and a grant from the University of the Basque Country, UPV/EHU, (University-Society US15/06 and US18/15). The authors thank for technical and human support provided by SCAB from SGIker of UPV/EHU.

Conflicts of Interest: The authors declare that they have no conflict of interest.

References

1. do Nascimento, A.B.; Fiates, G.M.; Dos Anjos, A.; Teixeira, E. Analysis of ingredient lists of commercially available gluten-free and gluten-containing food products using the text mining technique. *Int. J. Food Sci. Nutr.* **2013**, *64*, 217–222. [CrossRef] [PubMed]
2. Missbach, B.; Schwingshackl, L.; Billmann, A.; Mystek, A.; Hickelsberger, M.; Bauer, G.; König, J. Gluten-free food database: The nutritional quality and cost of packaged gluten-free foods. *PeerJ* **2015**, *3*, e1337. [CrossRef] [PubMed]
3. Wu, J.H.; Neal, B.; Trevena, H.; Crino, M.; Stuart-Smith, W.; Faulkner-Hogg, K.; Yu Louie, J.C.; Dunford, E. Are gluten-free foods healthier than non-gluten-free foods? An evaluation of supermarket products in Australia. *Br. J. Nutr.* **2015**, *114*, 448–454. [CrossRef] [PubMed]
4. Miranda, J.; Lasa, A.; Bustamante, M.A.; Churruca, I.; Simon, E. Nutritional differences between a gluten-free diet and a diet containing equivalent products with gluten. *Plant Foods Hum. Nutr.* **2014**, *69*, 182–187. [CrossRef] [PubMed]
5. Mazzeo, T.; Cauzzi, S.; Brighenti, F.; Pellegrini, N. The development of a composition database of gluten-free products. *Public Health Nutr.* **2015**, *18*, 1353–1357. [CrossRef] [PubMed]
6. Staudacher, H.M.; Gibson, P.R. How healthy is a gluten-free diet? *Br. J. Nutr.* **2015**, *114*, 1539–1541. [CrossRef]
7. Gliszczyńska-Świgło, A.; Klimczak, I.; Rybicka, I. Chemometric analysis of minerals in gluten-free products. *J. Sci. Food Agric.* **2018**, *98*, 3041–3048. [CrossRef]
8. Rybicka, I. The Handbook of Minerals on a Gluten-Free Diet. *Nutrients* **2018**, *10*, 1683. [CrossRef]
9. Yazynina, E.; Johansson, M.; Jägerstad, M.; Jastrebova, J. Low folate content in gluten-free cereal products and their main ingredients. *Food Chem.* **2008**, *111*, 236–242. [CrossRef]
10. Babio, N.; Alcázar, M.; Castillejo, G.; Recasens, M.; Martínez-Cerezo, F.; Gutiérrez-Pensado, V.; Masip, G.; Vaqué, C.; Vila-Martí, A.; Torres-Moreno, M.; et al. Patients with Celiac Disease Reported Higher Consumption of Added Sugar and Total Fat Than Healthy Individuals. *J. Pediatr. Gastroenterol. Nutr.* **2017**, *64*, 63–69. [CrossRef]
11. Barone, M.; Della Valle, N.; Rosania, R.; Facciorusso, A.; Trotta, A.; Cantatore, F.P.; Falco, S.; Pignatiello, S.; Viggiani, M.T.; Amoruso, A.; et al. A comparison of the nutritional status between adult celiac patients on a long-term, strictly gluten-free diet and healthy subjects. *Eur. J. Clin. Nutr.* **2016**, *70*, 23–27. [CrossRef]
12. Gobbetti, M.; Pontonio, E.; Filannino, P.; Rizzello, C.G.; De Angelis, M.; Di Cagno, R. How to improve the gluten-free diet: The state of the art from a food science perspective. *Food Res. Int.* **2018**, *110*, 22–32. [CrossRef]
13. Salazar Quero, J.C.; Espin Jaime, B.; Rodriguez Martinez, A.; Arguelles Martin, F.; Garcia Jimenez, R.; Rubio Murillo, M.; Pizarro Martin, A. Nutritional assessment of gluten-free diet. Is gluten-free diet deficient in some nutrient? *An. Pediatr.* **2015**, *83*, 33–39. [CrossRef]
14. Botero-López, J.E.; Araya, M.; Parada, A.; Méndez, M.A.; Pizarro, F.; Espinosa, N.; Canales, P.; Alarcón, T. Micronutrient deficiencies in patients with typical and atypical celiac disease. *J. Pediatr. Gastroenterol. Nutr.* **2011**, *53*, 265–270. [CrossRef]
15. Pellegrini, N.; Agostoni, C. Nutritional aspects of gluten-free products. *J. Sci. Food Agric.* **2015**, *95*, 2380–2385. [CrossRef]
16. Matos Segura, M.E.; Rosell, C.M. Chemical Composition and Starch Digestibility of Different Gluten-free Breads. *Plant Foods Hum. Nutr.* **2011**, *66*, 224. [CrossRef]
17. Schoenlechner, R.; Drausinger, J.; Ottenschlaeger, V.; Jurackova, K.; Berghofer, E. Functional Properties of Gluten-Free Pasta Produced from Amaranth, Quinoa and Buckwheat. *Plant Foods Hum. Nutr.* **2010**, *65*, 339–349. [CrossRef]
18. Hager, A.S.; Wolter, A.; Jacob, F.; Zannini, E.; Arendt, E.K. Nutritional properties and ultra-structure of commercial gluten free flours from different botanical sources compared to wheat flours. *J. Cereal Sci.* **2012**, *56*, 239–247. [CrossRef]
19. Alvarez-Jubete, L.; Arendt, E.K.; Gallagher, E. Nutritive value and chemical composition of pseudocereals as gluten-free ingredients. *Int. J. Food Sci. Nutr.* **2009**, *60*, 240–257. [CrossRef]

20. AESAN. Base de Datos Española de Composición De Alimentos (BEDCA) v 1.0. Available online: http://www.bedca.net/ (accessed on 30 July 2019).
21. Horwitz, W.; Latimer, G.W. *AOAC International. Official Methods of Analysis of AOAC International*, 18th ed.; (Revision 3, 2005); AOAC International: Gaithersburg, MD, USA, 2010.
22. Roe, M.; Pinchen, H.; Church, S.; Finglas, P. McCance and Widdowson's the Composition of Foods Seventh Summary Edition and updated Composition of Foods Integrated Dataset. *Nutr. Bull.* **2015**, *40*, 36–39. [CrossRef]
23. AUSNUT 2011-13 Food Nutrient Database. Available online: http://www.foodstandards.gov.au/science/monitoringnutrients/ausnut/ausnutdatafiles/Pages/foodnutrient.aspx (accessed on 30 July 2019).
24. USDA Food Composition Databases. Available online: https://ndb.nal.usda.gov/ndb/ (accessed on 30 July 2019).
25. Zuccotti, G.; Fabiano, V.; Dilillo, D.; Picca, M.; Cravidi, C.; Brambilla, P. Intakes of nutrients in Italian children with celiac disease and the role of commercially available gluten-free products. *J. Hum. Nutr. Diet.* **2013**, *26*, 436–444. [CrossRef]
26. Larretxi, I.; Simon, E.; Benjumea, L.; Miranda, J.; Bustamante, M.A.; Lasa, A.; Eizaguirre, F.J.; Churruca, I. Gluten-free-rendered products contribute to imbalanced diets in children and adolescents with celiac disease. *Eur. J. Nutr.* **2019**, *58*, 775–783. [CrossRef]
27. FESNAD. Dietary reference intakes (DRI) for spanish population, 2010. *Act. Diet.* **2010**, *14*, 196–197. [CrossRef]
28. Matos, M.E.; Rosell, C.M. Understanding gluten-free dough for reaching breads with physical quality and nutritional balance. *J. Sci. Food Agric.* **2015**, *95*, 653–661. [CrossRef]
29. Regulation (EU) No 1169/2011 of the European Parliament and of the Council of 25 October 2011 on the provision of food information to consumers, amending Regulations (EC) No 1924/2006 and (EC) No 1925/2006 of the European Parliament and of the Council, and repealing Commission Directive 87/250/EEC, Council Directive 90/496/EEC, Commission Directive 1999/10/EC, Directive 2000/13/EC of the European Parliament and of the Council, Commission Directives 2002/67/EC and 2008/5/EC and Commission Regulation (EC) No 608/2004 (Text with EEA relevance). Available online: http://data.europa.eu/eli/reg/2011/1169/2018-01-01 (accessed on 30 July 2019).
30. Lee, A.R.; Wolf, R.L.; Lebwohl, B.; Ciaccio, E.J.; Green, P.H.R. Persistent Economic Burden of the Gluten Free Diet. *Nutrients* **2019**, *11*, 399. [CrossRef]
31. Thompson, T. Thiamin, riboflavin, and niacin contents of the gluten-free diet: Is there cause for concern? *J. Am. Diet. Assoc.* **1999**, *99*, 858–862. [CrossRef]
32. Krupa-Kozak, U.; Wronkowska, M.; Soral-ŚMietaNa, M. Effect of Buckwheat Flour on Microelements and Proteins Contents in Gluten-Free Bread. *Czech J. Food Sci.* **2011**, *2*, 103–108. [CrossRef]
33. Rybicka, I.; Gliszczyńska-Świgło, A. Minerals in grain gluten-free products. The content of calcium, potassium, magnesium, sodium, copper, iron, manganese, and zinc. *J. Food Compos. Anal.* **2017**, *59*, 61–67. [CrossRef]
34. Rybicka, I.; Gliszczynska-Swiglo, A. Gluten-Free Flours from Different Raw Materials as the Source of Vitamin B. *J. Nutr. Sci. Vitaminol* **2017**, *63*, 125–132. [CrossRef]
35. Mariani, P.; Viti, M.G.; Montuori, M.; La Vecchia, A.; Cipolletta, E.; Calvani, L.; Bonamico, M. The gluten-free diet: A nutritional risk factor for adolescents with celiac disease? *J. Pediatr. Gastroenterol. Nutr.* **1998**, *27*, 519–523. [CrossRef]
36. Ohlund, K.; Olsson, C.; Hernell, O.; Ohlund, I. Dietary shortcomings in children on a gluten-free diet. *J. Hum. Nutr. Diet.* **2010**, *23*, 294–300. [CrossRef]
37. Food and Drug Regulations, Government of Canada. 2019. Available online: https://laws-lois.justice.gc.ca/eng/regulations/C.R.C.,_c._870/ (accessed on 30 July 2019).
38. Questions and Answers on FDA's Fortification Policy-Guidance for Industry. Available online: https://www.fda.gov/media/94563/download (accessed on 15 July 2019).
39. Melini, V.; Melini, F. Gluten-Free Diet: Gaps and Needs for a Healthier Diet. *Nutrients* **2019**, *11*, 170. [CrossRef]
40. Harper, J.W.; Holleran, S.F.; Ramakrishnan, R.; Bhagat, G.; Green, P.H. Anemia in celiac disease is multifactorial in etiology. *Am. J. Hematol.* **2007**, *82*, 996–1000. [CrossRef]

41. Olechnowicz, J.; Tinkov, A.; Skalny, A.; Suliburska, J. Zinc status is associated with inflammation, oxidative stress, lipid, and glucose metabolism. *J. Physiol. Sci.* **2018**, *68*, 19–31. [CrossRef]
42. Suliburska, J.; Krejpcio, Z.; Reguła, J.; Grochowicz, A. Evaluation of the content and the potential bioavailability of minerals from gluten-free products. *Acta Sci. Pol. Technol. Aliment.* **2013**, *12*, 75–79.
43. Caruso, R.; Pallone, F.; Stasi, E.; Romeo, S.; Monteleone, G. Appropriate nutrient supplementation in celiac disease. *Ann. Med.* **2013**, *45*, 522–531. [CrossRef]
44. Dennis, M.; Lee, A.R.; McCarthy, T. Nutritional Considerations of the Gluten-Free Diet. *Gastroenterol. Clin. N. Am.* **2019**, *48*, 53–72. [CrossRef]

© 2019 by the authors. Licensee MDPI, Basel, Switzerland. This article is an open access article distributed under the terms and conditions of the Creative Commons Attribution (CC BY) license (http://creativecommons.org/licenses/by/4.0/).

foods

MDPI

Article

Lipids and Fatty Acids in Italian Durum Wheat (*Triticum durum* Desf.) Cultivars

Valentina Narducci, Enrico Finotti, Vincenzo Galli and Marina Carcea *

Research Centre for Food and Nutrition, Council for Agricultural Research and Economics (CREA), Via Ardeatina 546, 00178 Rome, Italy; valentina.narducci@crea.gov.it (V.N.); enrico.finotti@crea.gov.it (E.F.); vincenzo.galli@crea.gov.it (V.G.)

* Correspondence: marina.carcea@crea.gov.it; Tel.: +30-06-5149-4638

Received: 22 May 2019; Accepted: 19 June 2019; Published: 21 June 2019

Abstract: The level of variation in lipids and their fatty acids was determined in the grains of 10 popular durum wheat cultivars commercially grown in Central and Southern Italy. Samples were harvested for two consecutive years to account for differences due to changes in climatic conditions. Total fat content was determined by means of the International Association of Cereal Science and Technology (ICC) Standard Method No. 136, whereas the fatty acid profile was determined by gas chromatography. Total lipid content ranged from 2.97% to 3.54% dry basis (d.b.) in the year 2010 and from 3.10% to 3.50% d.b. in the year 2011, and the average value was 3.22% d.b. considering both years together. Six main fatty acids were detected in all samples in order of decreasing amounts: linoleic (C18:2) > palmitic (C16:0) ≈ oleic (C18:1) > linolenic (C18:3) > stearic (C18:0) > palmitoleic (C16:1). Significant variations in the levels of single acids between two years were observed for three samples. These results will be very useful in the updating of food composition databases in general and will help authorities to set proper quality standards for wholegrain flours and products where the germ should be preserved, considering also the recent interest of industry and consumers for these kinds of products.

Keywords: durum wheat; fatty acids; grain; kernel; lipids

1. Introduction

Durum wheat (*Triticum durum* Desf.) kernels contain about 2.4–3.8% dry basis (d.b.) of lipids [1]. Roughly two thirds (66%) of them are contained in the germ, 15% are in the bran (particularly in the aleuronic layer), and about 20% are distributed in the endosperm, partly within the starch granules. From a chemical point of view, the most abundant fraction is composed by nonpolar lipids, which are mainly storage acylglycerols. Phospholipids, glycolipids and other classes are present in lesser amounts. The fatty acids of wheat lipids are mostly unsaturated (C18:2, C18:1, C18:3 and C16:1) and two of them are essential (linoleic and linolenic). This increases the value of wheat lipids for human nutrition, because essential fatty acids are precursors of important classes of biomolecules in the human body (like prostaglandins and membrane phospholipids) and are involved in metabolic processes like regulation of blood lipid levels, particularly cholesterol [1–3].

Lipid content, lipid classes and fatty acid levels in wheat kernels depend on a set of factors, some of which are genetic, such as species and variety [4], whereas others depend on the environment and are related to pedoclimatic conditions, agronomic practices and maturity level [1,4,5]. For example, durum wheat and hard red wheat generally have a higher lipid content than soft white wheat and the levels of fatty acids are different in durum and in soft wheat. In regard to climatic conditions, it has been seen that cold weather favors an increase of lipid content in wheat and a higher degree of unsaturation in fatty acids due to the need for membrane fluidification [6]. Other kinds of biotic and abiotic stresses can influence the level of saturated and unsaturated fatty acids in plants [7].

Moreover, different extraction and analytical methods can also account for the differences found in the literature [1,8]. Notwithstanding the number of samples analyzed, we can assume that data about fatty acid levels in durum wheat are abundant in the literature, but it is difficult to have a clear idea of their content and to make comparisons for a number of reasons: (i) different authors report fatty acids as percentage, alternatively referring to: (1) total lipids, (2) total fatty acids, or (3) kernel weight (in addition, some authors analyze germ oil and others analyze whole kernels); (ii) authors interested in statistic elaborations (e.g., in order to investigate variation factors or to look for discriminating parameters) often report charts and graphs rather than tables of data; (iii) cultivars are different in different countries and new ones are constantly bred; and (iv) databases do not always report the sample numerosity and the standard variation of the means.

In this work, the content and level of variation in lipids and of their fatty acids in the durum wheat kernels commercially grown in Italy (where durum wheat is an important cereal crop mainly used for pasta manufacturing) were assessed. For this reason, we selected 10 cultivars amongst the most commonly grown for pasta making. Samples were collected in several locations of Central and Southern Italy to account, at least partially, for differences due to different pedoclimatic environments; Southern Italy is characterized by milder winters and warmer springs and summers with respect to Central Italy, however both areas are considered highly suitable for durum wheat cultivation. Moreover, crops from two consecutive years were collected from the same fields.

The knowledge generated by this research will be very useful in the updating of food composition databases in general and will help authorities in setting proper quality standards for wholegrain flours and products where the germ should be preserved, considering also the recent interest of industry and consumers for these kinds of products and the lack, in several cases, of specific legislation.

2. Materials and Methods

2.1. Samples and Sample Preparation

Representative samples of durum wheat grains, belonging to 10 cultivars selected amongst the most frequently grown in Italy, were collected at harvest for two consecutive years (2010–2011) in 10 different locations of Central and Southern Italy (Table 1). Eight samples came from the Central regions of Italy (Tuscany and Marche) whereas twelve were from different locations in the Sicilian region, in the South. All locations belong to the area traditionally dedicated to durum wheat cultivation in Italy.

Table 1. Durum wheat sample specifications: cultivar, region and location.

Cultivar	Region	Location
Ancomarzio	Tuscany (Central Italy)	Siena (SI)
Creso	Tuscany (Central Italy)	Pisa (PI)
Dylan	Marche (Central Italy)	Macerata (MC)
Rusticano	Marche (Central Italy)	Ancona (AN)
Bronte	Sicily (Southern Italy)	Palermo (PA)
Ciccio	Sicily (Southern Italy)	Enna (EN)
Duilio	Sicily (Southern Italy)	Trapani (TP)
Iride	Sicily (Southern Italy)	Agrigento (AG)
K26	Sicily (Southern Italy)	Enna (EN)
Simeto	Sicily (Southern Italy)	Catania (CT)

Durum wheat in Italy is grown under rain-fed production: it is planted in late autumn or early winter and harvested in early summer, which often leads to limited rainfall and high temperatures, resulting in water stress during grain filling. Crop rotation and balanced nutrient management (mainly nitrogen and phosphorus, pre-sowing and topdressing fertilization) are practiced to ensure that the crop produces the greatest possible high-quality yield with the moisture that is available. The main climate factors influencing durum wheat crop quality are rainfall and temperature during the growing season. Data on these two factors of the years 2009–2011 in Central and Southern Italy can be found in the reports by the Italian High Institute for Environmental Protection and Research (ISPRA,) [9–11].

Fifty grams of each cleaned sample were milled by means of a Cyclotec laboratory mill (Foss-Tecator, Hillerød, Denmark) equipped with a 0.5 mm screen, to obtain wholemeal flours that were used for the subsequent analyses.

2.2. Chemicals

Chloroform, ethyl alcohol (96% *w/w*), methanol, *n*-hexane, formic acid (99% *w/w*) hydrochloric acid (37% *w/w*) and anhydrous sodium sulphate were of analytical grade and were purchased from Carlo Erba (Milan, Italy). Boron trifluoride (approximately 10% *w/w* in methanol for gas chromatography (GC) derivatization) was purchased from Sigma-Aldrich (St. Louis, MO, USA). Fatty acid standards (C16:0, C16:1, C17:0, C18:0, C18:1, C18:2, C18:3) were also purchased from Sigma-Aldrich.

2.3. Analyses

Moisture of wholemeal flours was determined by oven drying at 130 °C according to the ICC Standard No. 110/1 [12].

Total fat was determined by hydrolysis in formic acid and hydrochloric acid at 75 °C reflux for 20 min followed by extraction in hexane and evaporation, according to the ICC Standard No. 136 [12].

The fatty acid profile was determined by gas chromatography (GC). About 5 g of wholemeal flour (in duplicate) was introduced in a Corning tube and suspended in 10 mL of chloroform–methanol 2:1 acidulated with 6 N HCl. A magnetic bar was added, and the tube was left to extract overnight at room temperature on a magnetic stirrer. The mixture was filtered through Whatman Grade 1 (1–11 μm) filter paper into an oven dried flask, then the solvent was evaporated by nitrogen flux followed by oven drying at 30 °C. The contents of the flask were re-dissolved in chloroform–methanol 2:1 to a volume of exactly 10 mL, then an aliquot was derivatized according to Zweig and Sherma [13] as follows: 100 μL of this solution was introduced into a Corning tube containing 3 mL of methanol and a few boiling stones, then 0.5 mL of BF_3–methanol (10% *w/w*) was added and the tube caps were loosely screwed. The tubes were put onto a heating plate in a water bath and left to gently reflux at 72 °C for 30 min. Following this, the reaction was quenched with 2 mL of water, then the mixture was cooled to room temperature and extracted three times with 3 mL of *n*-hexane. The hexane extracts were reunited into a vial and finally the hexane was evaporated by nitrogen flux. The vial was stored under nitrogen at −18 °C for a few days. Immediately prior to GC analysis, the contents of the vial were re-dissolved in 300 μL of hexane and 2 μL were injected. The GC instrument was an HP 5890 equipped with a Supelco (Sigma-Aldrich, St. Louis, MO, USA) SPB®-PUFA (poly unsaturated fatty acids) column of 30 m length and a flame ionization detector (F.I.D.). The instrumental analysis was run according to Finotti et al. [14]: 50 °C for 1 min, ramp of 10 °C/min until 160 °C, stay at 160 °C for 1 min, ramp of 2 °C/min until 240 °C. The detected peaks were individuated by comparison with chromatograms of standards (C16:0, C16:1, C18:0, C18:1, C18:2, C18:3) and quantified by using C17:0 as an internal standard.

2.4. Statistics

The Shapiro–Wilks normality test, *F*-test for homogeneity of variance, Student's *t*-test and Friedman test followed by Wilcoxon pairwise comparisons were performed by means of the PAleontological STatistics (PAST) statistical package [15]. Two-way ANOVA followed by Tukey's test (only in cases with a normal variable and homogeneous variances) and box-plots were performed by means of StatSoft Statistica 8.0 (TIBCO Software, Palo Alto, CA, USA). Calculations were performed by means of Microsoft Excel (Redmond, Washington State, USA).

3. Results

3.1. Total Lipids

Total lipids ranged from 2.97% to 3.54% d.b. in the year 2010 and from 3.10% to 3.50% d.b. in the year 2011, and the average value was 3.22% d.b. considering both years together (Table 2). The moisture content of grains ranged between 10.5% and 12.3% and the average was 11.4% (Table 2). Total lipid content was strongly dependent on the combination of cultivar (cv)/growing site ($p < 0.01$) and to a minor extent on the growing year ($p < 0.05$), whereas the interaction cv/site × year was not a statistically significant factor of variation. In any case, differences were very small: up to 0.57 between samples of different cultivars and up to 0.18 between years for samples of a same cv/site (Table 2). Differences between years for samples of the same cv/site were not significant. The total lipid values found in this study are in line with those reported by the USDA National Nutrient Database (2.8 g/100 g d.b. for product N. 20076 "wheat, durum", mean of 18 samples, standard error 0.060) and by the Italian food composition tables (3.3 g/100 g d.b. for "durum wheat") compiled by the Italian National Institute for Research on Food and Nutrition (INRAN) [16,17]. If we take into account the geographical separation into Central and Southern Italy, we can say that the average total lipid values for all samples were 3.24% and 3.21% d.b. respectively, whereas the range of values was 2.97–3.54% for Central Italy and 3.09–3.41% d.b. for Southern Italy.

Table 2. Moisture and total lipids in the grains of 10 Italian durum wheat cultivars grown in different locations for two consecutive years.

Cultivar and Location		Moisture (g/100 g)		Total Lipids (g/100g d.b.)			
		2010	2011	2010	2011	Difference 2011–2010	
Central Italy	Ancomarzio SI	11.2	11.0	3.11 ^{ef}	3.25 ^{cde}	0.14	ns
	Creso PI	11.8	11.4	3.24 ^{cde}	3.25 ^{cde}	0.01	ns
	Dylan MC	11.7	11.7	3.54 ^{a}	3.50 ^{ab}	−0.04	ns
	Rusticano AN	12.3	12.0	2.97 ^{f}	3.10 ^{ef}	0.13	ns
Southern Italy	Bronte PA	11.3	11.1	3.09 ^{ef}	3.28 ^{bcde}	0.18	ns
	Ciccio EN	11.5	10.8	3.39 ^{abcd}	3.41 ^{abc}	0.02	ns
	Duilio TP	11.2	11.4	3.11 ^{ef}	3.15 ^{def}	0.05	ns
	Iride AG	11.0	10.5	3.31 ^{abcde}	3.24 ^{cde}	−0.07	ns
	K26 EN	11.3	11.0	3.10 ^{ef}	3.24 ^{cde}	0.14	ns
	Simeto CT	11.7	11.6	3.13 ^{ef}	3.10 ^{ef}	−0.02	ns

^{abcdef}: different letters correspond to significant differences ($p < 0.05$) according to 2-way ANOVA and Tukey's test. ns: not significant.

3.2. Fatty Acid Profile

Six main fatty acids were detected in all samples, as expected. In order of decreasing amounts, they are: linoleic (C18:2) > palmitic (C16:0) ≈ oleic (C18:1) > linolenic (C18:3) > stearic (C18:0) > palmitoleic (C16:1). This can be clearly seen from the box plot elaboration reported for each separate year and for the two years together (Figure 1). This distribution did not change whether considering both years separately or together. Detailed data of fatty acids in all samples are reported in Table 3.

Linoleic acid (C18:2) was present in amounts ranging from 0.50–1.14 g/100 g d.b. throughout all samples, with a mean of 0.68 and a standard deviation (SD) of 0.16 (Table 3). For comparison, the USDA National Nutrient Database reports 1.04 g/100 g d.b. for product N. 20076 "wheat, durum" and the INRAN food composition tables report 1.36 g/100 g d.b. for durum wheat. Neither database reports any information on standard errors for all acids.

Palmitic (C16:0) and oleic (C18:1) acids were detected in equal amounts. Palmitic acid ranged from 0.17–0.36 g/100 g d.b., mean 0.24 (SD 0.04) and oleic acid ranged from 0.17–0.43 g/100 g d.b., mean 0.24 (SD 0.07). The USDA reports 0.51 g/100 g d.b. for palmitic acid and 0.40 g/100 g d.b. for oleic acid, whereas the INRAN database reports 0.47 g/100 g d.b. and 0.38 g/100 g d.b., respectively (Table 3).

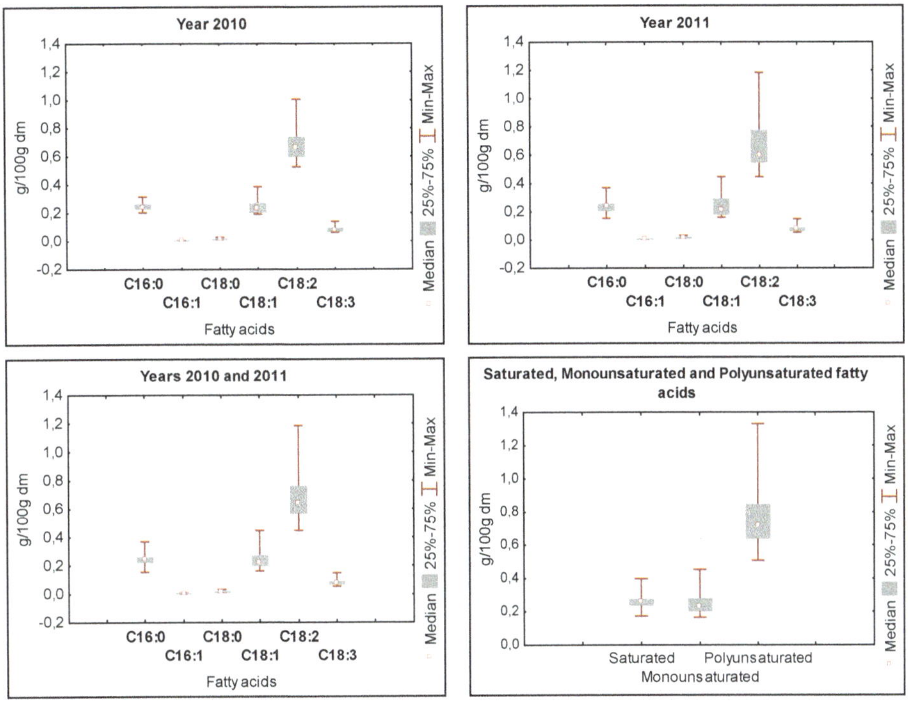

Figure 1. Box plot (percentiles) of fatty acids in samples of Italian durum wheat (10 cultivars, grown in the same location for two consecutive years).

Linolenic acid (C18:3) ranged from 0.06–0.14 g/100 g d.b., mean 0.08 (SD 0.02). The USDA and the INRAN databases report 0.05 g/100 g d.b. and 0.11 g/100 g d.b., respectively. Stearic acid (C18:0) ranged from 0.01–0.03 g/100 g d.b., mean 0.02 (SD 0.005). The USDA and the INRAN databases report, for this acid, 0.03 g/100 g d.b. and 0.02 g/100g d.b. respectively. Finally, palmitoleic acid (C16:1) was detected in very small amounts, ranging from 0.004–0.007 g/100 g d.b., mean 0.005 (SD 0.001). Both the USDA and INRAN databases report 0.01 g/100 g d.m. for this acid.

A series of *t*-tests, performed for each fatty acid on each pair of samples from the same cv/site between the two growing years, showed a significant difference between the years 2010 and 2011 in a few cases only, namely: all acids except C16:1 varied in Ancomarzio SI and Iride AG; only the acids C18:1, C18:2 and C18:3 varied in Ciccio EN (Table 3).

Table 3. Fatty acids in 10 Italian durum wheat cultivars, grown in different locations for two consecutive years (g/100 g sample, d.m.).

Sample	Total Lipids	C16:0	C16:1	C18:0	C18:1	C18:2	C18:3	Saturated	Monoun-Saturated	Polyun-Saturated	Ratio Unsaturated/ Saturated
Ancomarzio (SI) 2010	3.01	0.28	0.006	0.02	0.28	0.80	0.10	0.30	0.29	0.90	3.96
Ancomarzio (SI) 2011	3.03	0.23	0.004	0.01	0.18	0.54	0.06	0.24	0.18	0.60	3.24
		*			*	*	*				*
Creso (PI) 2010	3.02	0.23	0.004	0.02	0.24	0.68	0.07	0.25	0.24	0.75	3.97
Creso (PI) 2011	3.02	0.17	0.004	0.02	0.20	0.51	0.06	0.19	0.21	0.57	4.05
Dylan (MC) 2010	3.05	0.26	0.007	0.01	0.28	0.68	0.07	0.27	0.28	0.75	3.80
Dylan (MC) 2011	3.05	0.27	0.005	0.02	0.34	0.84	0.08	0.29	0.34	0.92	4.39
Rusticano (AN) 2010	3.00	0.23	0.005	0.02	0.22	0.65	0.08	0.25	0.23	0.72	3.87
Rusticano (AN) 2011	3.01	0.26	0.006	0.02	0.30	0.85	0.09	0.28	0.30	0.94	4.42
Bronte (PA) 2010	3.01	0.30	0.006	0.02	0.33	0.89	0.13	0.32	0.34	1.02	4.20
Bronte (PA) 2011	3.03	0.22	0.004	0.01	0.18	0.50	0.07	0.23	0.18	0.57	3.24
Ciccio (EN) 2010	3.04	0.24	0.004	0.02	0.25	0.73	0.09	0.26	0.25	0.83	4.11
Ciccio (EN) 2011	3.04	0.20	0.004	0.01	0.17	0.54	0.07	0.21	0.18	0.61	3.69
					*	**	**				
Duilio (TP) 2010	3.01	0.22	0.004	0.01	0.20	0.60	0.08	0.23	0.20	0.68	3.87
Duilio (TP) 2011	3.02	0.23	0.004	0.01	0.22	0.64	0.08	0.24	0.22	0.73	3.94
Iride (AG) 2010	3.03	0.23	0.007	0.01	0.22	0.59	0.07	0.24	0.22	0.66	3.62
Iride (AG) 2011	3.02	0.36	0.007	0.03	0.43	1.14	0.14	0.39	0.44	1.28	4.44
		**			**	**	**				*
K26 (EN) 2010	3.01	0.24	0.005	0.02	0.21	0.54	0.06	0.26	0.21	0.60	3.15
K26 (EN) 2011	3.02	0.24	0.005	0.01	0.22	0.59	0.07	0.26	0.23	0.66	3.48
							*				**
Simeto (CT) 2010	3.01	0.23	0.005	0.01	0.21	0.64	0.08	0.25	0.21	0.72	3.76
Simeto (CT) 2011	3.01	0.25	0.006	0.01	0.20	0.63	0.08	0.26	0.21	0.71	3.48
Max	3.00	0.36	0.007	0.03	0.43	1.14	0.14	0.39	0.44	1.28	4.44
Min	3.05	0.17	0.004	0.01	0.17	0.50	0.06	0.19	0.18	0.57	3.15
Mean	3.02	0.24	0.005	0.016	0.24	0.68	0.08	0.26	0.25	0.76	3.83
SD	0.014	0.04	0.001	0.005	0.07	0.16	0.02	0.04	0.07	0.18	0.38
Durum wheat USDA ‡	2.8	0.47	0.01	0.02	0.38	1.04	0.05	0.50	0.39	1.10	3.0
Durum wheat INRAN ‡	3.3	0.51	0.01	0.03	0.40	1.36	0.11	0.54	0.41	1.47	3.5

Asterisks indicate significant difference between the year 2010 and 2011, * $p < 0.05$, ** $p < 0.01$ (*t*-test). ‡: values on dry basis, calculated by the authors from original data in USDA and INRAN databases that are expressed on as-is basis.

3.3. *Saturated and Unsaturated Fatty Acids*

As expected, polyunsaturated fatty acids were preponderant over saturated and monounsaturated fatty acids in all samples ($p < 0.01$, Friedman test; see Figure 1), ranging from 0.57–1.28 g/100 g d.b. (Table 3). Total monounsaturated and total saturated, whose levels were roughly similar ($p < 0.01$), covered from 0.18–0.44 g/100 g d.b. and from 0.19–0.39 g/100 g d.b., respectively. The unsaturated/saturated ratio ranged from 3.15–4.44 g/100 g d.b. considering all samples, with a mean of 3.83 (SD 0.38). This mean is higher than that reported by USDA (3.0) and INRAN (3.5). A series of *t*-tests, performed on each pair of samples from the same cv/site grown in different years, showed a significant difference for the unsaturated/saturated ratio between years in only three cases (Ancomarzio SI, Iride AG and K26 EN) (Table 3).

4. Discussion

Total lipids were in line with the values reported by the USDA and the INRAN databases (nearer to the Italian value) and it was not possible to detect any difference between the geographical areas of Central and Northern Italy.

In regard to fatty acid composition, even if Bottari et al. in 1999 [18] observed the presence of more than 60 peaks by gas chromatography and mass spectrometry (GC-MS) and identified fatty acids with even numbers of carbon atoms from C12 to C30 as well as C15 and C17, the major fat components were saturated and unsaturated C16 and C18 and particularly C16:0, C18:1 and C18:2, which together represented around 90% of the total.

Actually, the USDA database (but not the INRAN one) and other works also report small amounts of C14:0 in durum wheat kernels (USDA 0.003 g/100 g fresh matter, corresponding to 0.0035 g/100 g d.b.). We did not detect this acid, as it was at the limit of detection of our method. There are publications reporting other fatty acids as well (i.e., C17, C20, C22 and C24), some in kernels (Beleggia et al. [5] who uses a GC-MS instrument) and others in germ oil [19,20]. However, only C16:0, C18:0, C18:1, C18:2 and C18:3 are constantly reported by all published works and are regarded as the most important ones in durum wheat, with others amounting to about 1–2% in total [1].

For all fatty acids except C18:3 and for total saturated, total monounsaturated and total polyunsaturated acids, the mean calculated for our samples was lower than the values reported by USDA and INRAN (roughly two thirds–half, $p < 0.01$ against a hypothetical value; see Table 3). However, the range of the detected values contained the reference values, except for C16:0 and for total saturated acids, for which the detected range extended entirely below the USDA and INRAN means. Neither database reports the standard deviations for fatty acids in durum wheat and only the USDA one reports sample numerosity (that is, 29); in this latter case, a certain width around the reported value can be supposed, but it is not quantified. On the contrary, for the unsaturated/saturated ratio, the range of the detected values extends entirely above the mean reported by USDA and contains that reported by INRAN. As a matter of fact, there are notable differences between the two references used. The INRAN values are equal to or higher than the USDA values for the considered variables, in particular for C18:0 (+50%), C18:2 (+31%), C18:3 (+120%), total polyunsaturated acids (+34%) and unsaturated/saturated ratio (+17%).

All the reported differences can be explained by the differences in genetic characteristics, pedoclimatic conditions, agronomical treatments and analytical procedure, as stated in the Introduction. In particular, Beleggia et al. [5] identified the interaction genotype × year × treatment as the main contributor to the variability of the fatty acid levels observed in 24 durum wheat samples, especially for linoleic, oleic and stearic acids. Armanino et al. [4] linked the fatty acid profile of 135 samples of durum wheat to the cultivar, the geographic origin and the harvest year. The variation in saturated and unsaturated fatty acids within the same variety is also associated with various kinds of biotic and abiotic stresses, like low or high temperature, salt, drought, pathogens and others [6,7].

Also, in our study, different conditions related to location and climatic factors can account for some of the observed variability in lipid parameters. In fact, from the ISPRA reports [9–11], we can briefly say that in both areas of Italy (Central and South), temperatures were similar in the first part of the two growing seasons (October–December 2009 and 2010), except the month of December which was warmer in 2009 than in 2010. In the second part of the growing season (January–June, particularly April–June), the Central area showed warmer temperatures in 2011 than in 2010. In regard to precipitation, the first growing season (2009) started with a lesser amount of rain in October–December with respect to the second one (2010) and continued with a higher amount of rain in the January–June period. This happened both in Central and Southern Italy.

5. Conclusions

This work contributes to the knowledge on the content and variability of total fats and of the main fatty acids in durum wheat kernels. The values obtained in this study are also compared with reference values from national and international databases. In this paper, the use of standard methods of analysis, statistical data (numerosity of samples, mean, standard errors) and the specification of all the elements that allow for conversion of results into different units of measure (g/100 g dry or wet sample, g/100 g fat matter) make this data very useful in the compilation of databases and easy to compare with other data. Moreover, updated data on lipids are needed to set proper quality standards for products such as wheat wholegrain flours and foods where the presence of germ is desirable.

Author Contributions: Conceptualization and supervision, M.C. and E.F.; investigation, E.F., V.N. and V.G.; data curation and writing—original draft preparation, V.N.; funding acquisition and writing—review and editing, M.C.

Funding: This research was supported by the Project PASTA-COUS financed by the Italian Ministry of Education, University and Research (MIUR) (Grant Number CUP H36J16000730001).

Acknowledgments: The authors wish to acknowledge Luigi Bartoli's technical assistance in sample preparation.

Conflicts of Interest: The authors declare no conflict of interest.

References

1. Lafiandra, D.; Masci, S.; Sissons, M.; Dornez, E.; Delcour, J.A.; Courtin, C.M.; Caboni, M.F. Kernel components of technological value. In *Durum Wheat Chemistry and Technology*, 2nd ed.; Sissons, M., Marchylo, B., Abecassis, J., et al., Eds.; AACC International Inc.: St. Paul, MN, USA, 2012; pp. 85–124.
2. Russo, G.L. Dietary *n*-6 and *n*-3 polyunsaturated fatty acids: From biochemistry to clinical implications in cardiovascular prevention. *Biochem. Pharmacol.* **2009**, *77*, 937–946. [CrossRef] [PubMed]
3. Mozaffarian, D.; Wu, J.H.Y. Omega 3 fatty acids and cardiovascular disease. *J. Am. Coll. Cardiol.* **2011**, *58*, 2047–2067. [CrossRef] [PubMed]
4. Armanino, C.; De Acutis, R.; Festa, M.R. Wheat lipids to discriminate species, varieties, geographical origins and crop years. *Anal. Chim. Acta* **2002**, *454*, 315–326. [CrossRef]
5. Beleggia, R.; Platani, C.; Nigro, F.; De Vita, P.; Cattivelli, L.; Papa, R. Effect of genotype, environment and genotype-by-environment interaction on metabolite profiling in durum wheat (*Triticum durum* Desf.) grain. *J. Cereal Sci.* **2013**, *57*, 183–192. [CrossRef]
6. Nejadsadeghi, L.; Maali-Amiri, R.; Zeinali, H.; Ramezanpour, S.; Sadeghzade, B. Membrane fatty acid compositions and cold-induced responses in tetraploid and hexaploid wheats. *Mol. Biol. Rep.* **2015**, *42*, 363–372. [CrossRef] [PubMed]
7. Upchurch, R.G. Fatty acid unsaturation, mobilization, and regulation in the response of plants to stress. *Biotechnol. Lett.* **2008**, *30*, 967–977. [CrossRef] [PubMed]
8. Rosicka-Kaczmarek, J.; Miśkiewicz, K.; Nebesny, E.; Makowski, B. Composition and functional properties of lipid components from selected cereal grains. In *Plant Lipids Science, Technology, Nutritional Value and Benefits to Human Health*; Budryn, G., Żyżelewicz, D., Eds.; Transworld Research Network: Kerala, India, 2015; pp. 119–145.
9. Istituto Superiore per la Protezione e la Ricerca Ambientale (ISPRA). Gli indicatori del clima in Italia nel 2009. Available online: http://www.isprambiente.gov.it/it/pubblicazioni/stato-dellambiente/gli-indicatori-del-clima-in-italia-nel-2009 (accessed on 14 June 2019).
10. Istituto Superiore per la Protezione e la Ricerca Ambientale (ISPRA). Gli indicatori del clima in Italia nel 2010. Available online: http://www.isprambiente.gov.it/it/pubblicazioni/stato-dellambiente/gli-indicatori-del-clima-in-italia-nel-2010-anno (accessed on 14 June 2019).
11. Istituto Superiore per la Protezione e la Ricerca Ambientale (ISPRA). Gli indicatori del clima in Italia nel 2011. Available online: http://www.isprambiente.gov.it/it/pubblicazioni/stato-dellambiente/gli-indicatori-del-clima-in-italia-nel-2011-anno-vii (accessed on 14 June 2019).
12. International Association for Cereal Science and Technology. *Standard Methods of the ICC*; The Association: Vienna, Austria, 2003.
13. Zweig, G.; Sherma, J. *Handbook of Chromatography*; CRC Press: New York, NY, USA, 1974; pp. 95–240.
14. Finotti, E.; Bersani, A.; Bersani, E. Total quality index for extra virgin olive oil. *J. Food Qual.* **2007**, *30*, 911–931. [CrossRef]
15. Hammer, Ų.; Harper, D.A.T.; Ryan, P.D. PAST: Paleontological statistics software package for education and data analysis. *Palaeontol. Electron.* **2001**, *4*, 1–9.
16. Istituto Nazionale di Ricerca per gli Alimenti e la Nutrizione (INRAN). Tabelle di composizione degli alimenti: 'frumento duro' (Food composition tables 'durum wheat'). Available online: http://nut.entecra.it/646/tabelle_di_composizione_degli_alimenti.html (accessed on 21 May 2019).

17. USDA National Nutrient Database for Standard Reference. Nutrient data for product 20076 'Wheat, durum'. Available online: https://ndb.nal.usda.gov/ndb/foods/show/20076?n1=%7BQv%3D1%7D&fgcd=&man=&lfacet=&count=&max=25&sort=default&qlookup=WHEAT+DURUM&offset=&format=Full&new=&measureby=&Qv=1&ds=&qt=&qp=&qa=&qn=&q=&ing= (accessed on 21 May 2019).
18. Bottari, E.; De Acutis, R.; Festa, M.R. On the lipid constituents of wheat of different species, variety, origin and crop year. *Ann. Chim.* **1999**, *89*, 849–862.
19. Zarroug, Y.; Mejri, J.; Dhawefi, N.; Ali, S.B.S.; El Felah, M.; Hassouna, M. Comparison of chemical composition of two durum wheat (*Triticum durum* L.) and bread wheat (*Triticum aestivum* L.) germ oils. *EKIN J. Crop Breed. Genet.* **2015**, *1*, 69–76.
20. Güven, M.; Kara, H.H. Some chemical and physical properties, fatty acid composition and bioactive compounds of wheat germ oils extracted from different wheat cultivars. *J. Agric. Sci.* **2016**, *22*, 433–443.

© 2019 by the authors. Licensee MDPI, Basel, Switzerland. This article is an open access article distributed under the terms and conditions of the Creative Commons Attribution (CC BY) license (http://creativecommons.org/licenses/by/4.0/).

MDPI
St. Alban-Anlage 66
4052 Basel
Switzerland
Tel. +41 61 683 77 34
Fax +41 61 302 89 18
www.mdpi.com

Foods Editorial Office
E-mail: foods@mdpi.com
www.mdpi.com/journal/foods

Milton Keynes UK
Ingram Content Group UK Ltd.
UKHW050702141123
432489UK00004B/173

9 783039 362103